Differential Geometric Methods in Mathematical Physics

MATHEMATICAL PHYSICS STUDIES

A SUPPLEMENTARY SERIES TO
LETTERS IN MATHEMATICAL PHYSICS

Editors:

J. C. CORTET, *Université de Dijon, France*
M. FLATO, *Université de Dijon, France*
M. GUENIN, *Institut de Physique Théorique, Geneva, Switzerland*
E. H. LIEB, *Princeton University, U.S.A.*
R. RACZKA, *Institute of Nuclear Research, Warsaw, Poland*

Editorial Board:

W. AMREIN, *Institut de Physique Théorique, Geneva, Switzerland*
H. ARAKI, *Kyoto University, Japan*
A. CONNES, *I.H.E.S., France*
L. FADDEEV, *Steklov Institute of Mathematics, Leningrad, U.S.S.R.*
J. FROHLICH, *I.H.E.S., France*
C. FRONSDAL, *UCLA, Los Angeles, U.S.A.*
I. M. GELFAND, *Moscow State University, U.S.S.R.*
A. JAFFE, *Harvard University, U.S.A.*
M. KAC, *The Rockefeller University, New York, U.S.A.*
A. A. KIRILLOV, *Moscow State University, U.S.S.R.*
A. LICHNEROWICZ, *Collège de France, France*
B. NAGEL, *K.T.H., Stockholm, Sweden*
J. NIEDERLE, *Institute of Physics CSAV, Prague, Czechoslovakia*
A. SALAM, *International Center for Theoretical Physics, Trieste, Italy*
W. SCHMID, *Harvard University, U.S.A.*
I. E. SEGAL, *M.I.T., U.S.A.*
J. SIMON, *Université de Dijon, France*
D. STERNHEIMER, *Collège de France, France*
I. T. TODOROV, *Institute of Nuclear Research, Sofia, Bulgaria*
S. ULAM, *University of Colorado, U.S.A.*

VOLUME 6

Differential Geometric Methods in Mathematical Physics

Edited by

S. STERNBERG
Harvard University and University of Tel Aviv

D. Reidel Publishing Company
A MEMBER OF THE KLUWER ACADEMIC PUBLISHERS GROUP
Dordrecht / Boston / Lancaster

Library of Congress Cataloging in Publication Data

CIP

Main entry under title:

Differential geometric methods in mathematical physics.
(Mathematical physics studies; v. 6)
Proceedings of the XI Annual Conference on Differential Geometric Methods in Mathematical Physics, held in Jerusalem, Aug. 5-11, 1982.
Includes index.
1. Geometry, Differential–Congresses. 2. Mathematical physics–Congresses. I. Sternberg, Shlomo. II. Conference on Differential Geometric Methods in Mathematical Physics (11th: 1982: Jerusalem) III. Series.
QC20.7.D52D52 1984 530.1'5636 84-8242
ISBN 90-277-1781-8

Published by D. Reidel Publishing Company
P.O. Box 17, 3300 AA Dordrecht, Holland

Sold and distributed in the U.S.A. and Canada
by Kluwer Academic Publishers,
190 Old Derby Street, Hingham, MA 02043, U.S.A.

In all other countries, sold and distributed
by Kluwer Academic Publishers Group,
P.O. Box 322, 3300 AH Dordrecht, Holland

Printed in The Netherlands

CONTENTS

PREFACE

The following pages represent the Proceedings of the XI Annual Conference on Differential Geometric Methods in Mathematical Physics which was held in Jerusalem from August 5 through 11, 1982 under the auspices of the Tel Aviv University and the Israel Academy of Sciences and Humanities.
In addition to the above mentioned institutions, partial financial support was received form the Bank Leumi LeIsrael Fund for International Conferences, the American Friends of the Tel Aviv Institute of Mathematical Sciences and the Mathematics and Physics Branch of the United States Army Research, Development and Standardization Group (UK).
We are grateful to all of these organizations for their financial support.

GAUGE THEORY AND NUCLEAR STRUCTURE

K. Bleuler
Institut für Theoretische Kernphysik
der Universität Bonn
Nußallee 14-16, D-5300 Bonn, West-Germany

I. INTRODUCTION

The recent, most impressive verification of the Salam-Weinberg theory of electro-weak interactions through the experimental discovery of the so-called intermediate bosons represents, at the same time, a success of the general gauge theoretical viewpoints in modern particle physics (quantum chromodynamics, QCD). This theory leads to a deeper and by far more natural interpretation of particle interaction and induces, as we shall see, also a profound change in our understanding of nuclear structure.

Here, just a very brief recapitulation of the main points: As fundamental particles one considers characteristic groups of fermions (denoted by the introduction of the color and flavor index) described by the Dirac equation (electrons, quarks, etc.) which interact through a characteristic coupling to

S. Sternberg (ed.), Differential Geometric Methods in Mathematical Physics, 1–22.

various gauge fields of bosonic type. The decisive point consists in the (practically unique) determination of the corresponding field equations and coupling terms (which constitute the very basis for the theoretical description of physical phenomenons) through a general invariance condition, namely the so-called local gauge principle: If the various groups of fermions are visualized as vectors in complex abstract linear spaces, the assumption of local gauge invariance states that the whole system has to be invariant under space-time dependent unitary rotations SU(N) within these new spaces.

It should be noted that there are various experimental facts which indicate the existence of such invariance groups in the global (space-time independent) case: First of all, the structure of the invariance group of the Hamiltonian of any bound system induces corresponding representations (which are related to measurable quantities) in the various eigen-spaces. An analysis of nuclear level schemes led, according to Heisenberg, to the invariance with respect to the so-called isospin (SU(2)) rotations which, enlarged to a general space-time dependence, were the origin of the well-known Yang-Mills theory, whereas a systematic ordering of hadron states suggested, according to Gell-Mann, the enlargement from SU(2) to SU(3) acting on the flavor components. The introduction of the 3-dimensional color space with the corresponding second SU(3) invariance was a consequence of the inner (quark) structure of hadrons in connection with the Pauli principle and special empirical properties of these systems.

The decisive mathematical step (considered for the first time by H. Weyl in an interesting, somewhat different form in general relativity) was, in particular in the case of color-SU(3), the transition from global to local, i.e. to space-time dependent transformations U(x) of the Dirac-spinor field $\psi(x)$ (the color indices are suppressed):

$$(1) \qquad \psi(x) = U(x)\psi'(x)$$

If introduced into the free Dirac-equation

$$(2) \qquad \gamma^\mu \frac{\partial}{\partial x^\mu} \psi + \kappa\psi = 0$$

an additional invariance distroying term of the form $\frac{\partial U}{\partial x^\mu} \psi'$ appears in the transformed equation which has to be compensated by the new interaction term with the 8-component gauge field A^a_μ (a = 1...8) which reads, if combined with the derivation occurring in (2) (summation again suppressed)

$$(3) \qquad D_\mu = \frac{\partial}{\partial x^\mu} + A^a_\mu \Gamma^a$$

where the Γ^a are the elements of the Lie-algebra of SU(3). In order to obtain a gauge invariant Lagrangian for the field equation for A^a_μ one forms out of the connection form given by (3) the corresponding curvature R

$$(4) \quad R_{\mu\nu} = F^a_{\mu\nu}\Gamma^a \quad \text{with} \quad F^a_{\mu\nu} = \frac{\partial A_\nu}{\partial x^\mu} - \frac{\partial A_\mu}{\partial x^\nu} + G^{abc}A^b_\mu A^c_\nu$$

where G represents the structure constants of the Lie--algebra of SU(3). The Lagrangian L is now given by (one chooses the most natural expression)

(5) $$L = {}^{a}F_{\mu\nu}\,{}^{a}F^{\mu\nu}$$

and leads, in view of the second, non-linear, term in the expression (4) for F, to a non-linear, i.e. self interaction field equation.

These very elementary developments are given here in order to show the perfect analogy to the transition from special to general relativity (formally corresponding to the transition from global to local invariance) where the partial derivation of a vector field v^{μ} has to be enlarged to the covariant derivation corresponding to (3):

(6) $$(D_{\mu}v)^{\nu} = \frac{\partial v^{\nu}}{\partial x^{\mu}} + \Gamma^{\nu}_{\mu\lambda}v^{\lambda}$$

thus introducing automatically the interaction with the gravitational field. Here, again, the most natural scalar, i.e. the curvature scalar R, formed out of the corresponding curvature tensor is introduced successfully in the Lagrangian of the gravitational field. At this point it should be remarked that the gauge transformation scheme as given above may be taken over (practically unchanged) into the framework of general relativity: In this case, however, the introduction of local gauge (fundamental for the whole theory) appears much more intuitive and practically compulsory.

Summarising, it is impressive to see that practically all basic laws of physics((Maxwell's equations are based on SU(1) and the Weinberg-Salam scheme on SU(1) × SU(2) are founded on very much the same geometric arguments although the physical meaning of the various quantities as well as the corresponding arguments for their introduction are very different. (The only exception constitutes the rather unpleasant "Higgs-term" in the electro-weak case). In this paper we are interested in the practical application of the color SU(3) gauge theory in order to determine the inner structure of the hadrons which represent an infinite sequence of relatively heavy, strongly interacting particles of bosonic as well as fermionic type. The best known examples are the π, ρ, ω-bosons and the nucleons together with various hyperons like Λ, Σ etc. in the fermionic case. In view of their apparently irregular physical properties (masses, magnetic moments, mutual interaction, etc.) it turned out, as suggested by Heisenberg already in 1946, to be most inappropriate to consider these particles to be elementary, i.e. to describe each of them by a separate field and their interaction by terms to be determined in a phenomenological way in each case. The decisive new idea is to visualize all hadrons (of fermionic as well as bosonic type) as systems of (colored) quarks bound by the corresponding gauge field A_μ^a acc. (3). From this viewpoint, in principle all characteristic individual properties of hadrons mentioned above are now to be described on the basis gauge theory which replaces the large number of individual fields and phenomenological interaction terms. This amounts, in fact, to an enormous concentration and systemization of physical laws. In view of the fact that the nucleus is definitely build on hadrons this program represents a fundamental change of conventional nuclear theory in which, so far, the nucleons (i.e. protons and neutrons) were considered to be elementary and in which the binding was determined through their (phenomenological) coupling to various boson fields. In view of the fact that both, nucleons as well as bosons, are just special hadrons with characteristic inner structure the entire nucleus appears to be a most complicated system of inter-correlated quarks. The scope of this paper is to give a survey (without any details) of our attempt to describe the nucleus from the outset as a heavy hadron, i.e. as a system of quarks bound by the corresponding gauge field without using the phenomenological concepts mentioned above. The first question to be answered is whether this rather revolutionary view-point might lead directly to the interpretation of a few characteristic, experimentally well-known nuclear properties. Before answering in a positive sense, I would like, however, to summarize the present (in fact unsatisfactory) situation of nuclear physics and to indicate various arguments in favor of this new comprehensive gauge theoretical viewpoint.

II. PRESENT STATUS OF NUCLEAR THEORY

Present day theory of matter is divided into three well-separated domains:

1) Atoms built on electrons bound by the Maxwellian field or, in language of quantized fields, by exchange of photons; space dimension 10^{-8} cm, characteristic energies 1 electron volt.
2) Nuclei built on nucleons (protons and neutrons) bound by bosonic fields (according Yukawa) or exchange of bosons; space dimension 10^{-12} cm, characteristic energies 1 MeV.

3) Nucleons built on quarks (3-valence quarks completed by so-called sea-quarks) bound by the gauge field or gluon exchange; space dimension 10^{-13} cm, characteristic energies 100 MeV.

This division is generally accepted and corresponds, at the same time, to an unfortunate separation between scientists. It is, however, important to compare the basic theoretical assumptions, the experimental verifications as well as the mutual influences to be expected. In spite of the apparent similarity of these three systems there are deep-rooted differences: First of all it is a striking fact that 1) and 2) are numerically well separated (factor 10^4 in spatial dimensions and 10^6 in energies) reducing thus mutual influences to a minimum, whereas the separation between 2) and 3) is by far less significant, (10^1 in space dimension and 10^2 in energies, i.e., values which practically forbid a clear-cut separation). On the other hand, the binding fields in 1) and 3) are genuine gauge fields whereas the binding in 2) occurs through the exchange of bosons, i.e. particles with characteristic inner structure whose interaction must be determined in a phenomenological way. In this connection it is interesting to realize that, while the theoretical predictions in 1) are most satisfactory and in some cases of extreme accuracy (up to 10 or 11 decimals in the verification of radiative corrections), all results from conventional nuclear theory based on boson exchange lead at most to accuracies of 1 - 10%. At the same time it turned out that a fully satisfactory treatment of boson-exchange was, in fact, not possible. The major problem stems, however, from the fact that the systems 2) and 3) cannot be considered - in view of the relative dimensions given above - as separated structures: The relatively large dimensions of the nucleons and the corresponding low excitation energies lead within nuclear matter to an appreciable spatial overlap as well as to a relatively strong internal excitation of the nucleons (related to their quark structure). It is, in this connection, interesting to observe that a recent experiment about the measurement of quark distribution in nuclear matter, as compared to the corresponding distribution in free nucleons, shows explicitly that the nucleons immersed into nuclear matter exhibit a characteristic change of their inner structure[7)]. These facts lead naturally to the idea (mentioned in sect. I) to treat the nucleus from the outset as a system of quarks, i.e., on the level of nucleons or any other hadrons. In other words: The separation between the domains 2) and 3) should vanish.

Under these circumstances it becomes of major interest to solve, at least approximately, the many-quark system with a mutual interaction determined through the gauge field. This constitutes, however, an enormous, for the moment unsolvable, task. The difficulties are mainly due to the non-linearity (and the strength of the interaction) of the gauge field. Therefore, the existing theories are based so far to a large extent on simplifying assumptions. First of all one has to take into account the experimentally well-established confinement (of quarks and gluons) which states that no single quarks nor single gluons are allowed to escape from bound systems. In order to satisfy this requirement one introduces a rigorous boundary condition on the surface of the system which might be visualized as the effect of an infinitely high potential well. It is hoped - and to some extent verified by the so-called lattice approximation to gauge theory - that the confinement will be eventually understood as a consequence of the characteristic non-linearity of the gauge field. (There is a certain similarity to the formation of "black holes" in gravitational theory). In a second step a direct interaction between the quarks in principle due to the quantised gauge field is introduced in a simplified way maintaining, however, a global SU(3) invariance. In addition, one has to meet the empirically well established supplementary condition about vanishing total color: In view of the gauge invariance of the system which ascribes a characteristic representation of SU(3) to every eigen-space this condition selects the singlet states, i.e., the states corresponding to the 1-dimensional representation. As a consequence color becomes so to speak unobservable. With the help of these approximations (called "bag models")[1)] it was, however, possible to describe a large number of characteristic empirical properties of light hadrons, in particular of the nucleons with their typical charge distributions, magnetic moments, excited states a.s.o.

The method adopted in this work mainly consists in extending this approximation scheme (i.e., the bag model) to the entire nuclear structure. In doing so, one has to meet, however, some rather restrictive conditions: It is well-known that conventional nuclear theory, as described and critizised above, led, nevertheless, in certain cases to a successful interpretation of various characteristic empirical nuclear properties. In this work we consider the special data related to nuclear shell-structure[2)] which are very similar to the ones known in the atomic case: The well-known approximation scheme consists in introducing an average spherical potential and the determination of the corresponding degenerate single particle levels which form the various shells. The fundamental state is given by a compact occupation of the lowest levels according to the Pauli principle. This procedure leads to the experimentally well established characteristic steps (or discontinuities) of separation energies if one compares a nucleus with a competely filled shell (magic state) to a system with an additional fermion (which in this case has to occupy the next higher energy level). Detailed calculation within the conventional scheme (using as a refinement the so-called pairing approximation in order to take care of the additional direct interaction between the nucleons) leads to a rather good agreement with the empirical values[3)]. One, therefore, has the impression that the conventional picture represents the only possible scheme to reproduce these characteristic nuclear properties. On the other hand, the existence of the "sharp" energy steps following the magic numbers appears to be in contradiction with the fact,mentioned above,about the (partial) dissolution of nucleons immersed into nuclear matter: This effect prevents us to apply the Pauli principle which, on the other hand, is definitely needed in order to obtain the typical effects of filled shells or 'magic numbers'. In other words: Characteristic shell structure effects can only be obtained from ideal fermions, i.e., particles which remain unchanged relative to their inner structure within the bound system.

The purpose of our work is to show that these ideal fermions are, in fact, represented by the quarks and not by the nucleons. In other words, it will be seen that the system of bound quarks representing the entire nucleus reproduces the characteristic shell effects in a much more natural and even simpler way.

III. NUCLEAR SHELL STRUCTURE FROM A QUARK THEORETICAL VIEWPOINT

We thus consider a spherical heavier nucleus as a bound system of quarks treated with very much the same approximation scheme as used for the description of light hadrons, in particular free nucleons[1)]. This amounts to introduce, first of all, a confining condition on the spherical surface of the nucleus and to consider within the framework of a first approximation the quarks as independent Dirac particles with three color components (index $k = 1...3$) and two isospin states for "up" and "down" quarks (index $\tau = 1,2$) corresponding to the first flavor components. This yields the following eigen-value problem for the so-called single particle wave function (in conventional notation):

$$(1) \qquad (\vec{\alpha}\,\frac{\partial}{\partial\vec{x}} + \beta m)\,\psi_{k\tau} = \varepsilon\psi_{k\tau}$$

with the boundary condition for $|\vec{x}| = R$:

$$\frac{\vec{\alpha}\vec{x}}{|\vec{x}|}\,\psi_{k\tau} = i\beta\psi_{k\tau}\ . \tag{2}$$

The radius R of our surface sphere is to be determined through a special equilibrium condition, used already in the original bag models, saying that a hypothetical outer pressure B compensates the inner pressure of the quarks due to the volume (radius)-dependence of their energy eigen-values. (It is generally assumed that B constitutes a characteristic property of the degenerate quantum chromodynamical vacuum state, a fact which has been, to some extent, checked with the help of the so--called lattice approximation scheme; a satisfactory treatment of this fundamental and far-reaching quantum field theoretical problem is, however, not yet in sight). This extremely simple prescription taken over from earlier work about the quark structure of nucleons (i.e., the well-known bag models[1)]) yields, strangely enough, rather satisfactory results: With appropriate values for the pressure B and the mass m of the quarks one obtains immediately the nuclear radius law, a reasonable value for the nuclear binding and, what is most important, a single particle level scheme (i.e. ε-values according to (1)) which corresponds favorably with the ones occurring in the conventional nuclear shell model[2)]. In contrast to this former (non-relativistic) theory, where a more or less phenomenological spin-orbit term had to be introduced, the level scheme is given here directly by the relativistic Dirac eigen-value problem according (1) and (2). In this way the experimentally known shell closures with 3N-quarks instead of N-nucleons are obtained in a most natural way. (The additional condition of vanishing color is in this case automatically satisfied). In this paper we call the 2(2j+1) states which belong to an eigen-value with given j and parity a nuclear shell; as we neglect the Coulomb--energy the two isospin states are degenerate and belong to the same shell.

The main task consists, however, in the treatment of the open shells and the interpretation of the corresponding empirical properties of nuclei which are represented in a satisfactory way within the conventional theory. (There occur characteristic degeneracies and assignments in agreement with experimental findings). In the new quark theoretical case, however, the open shells behave at first sight in a perfectly different way: They exhibit, in view of the additional color degrees of freedom, much larger degeneracies as well as experimentally not observed assignments.

The decisive step of this work consists in the introduction of a direct interaction between the quarks which, in principle, is due to the coupling to the gauge field. So far, several explicit(approximate) expressions have been given; they contain always a certain number of unknown parameters. If, however, one of these expressions is used in connection with the well-known pairing approximation method for N-fermion systems <u>and</u> the fundamental SU(3) invariance we are left with a single strength parameter g for each shell (we consider "pairing" only between state in a given shell). This well-known approximation scheme, i.e., the so-called BCS-theory, plays an important role also in other domains: It constitutes the basis of the theory of super-conductivity and was, at the same time, used most successfully in conventional nuclear structure[3]. For the purpose of our quark system a slight generalization had to be developed. In order to write this "pairing interaction term" explicitly we introduce (as usual) the second quantization scheme based on the definition of the following well-known operator algebra (* represents the hermition conjugate)

$$(3)\quad [a_r a_s^*]_+ = \delta_{rs} I, \qquad [a_r a_s]_+ = [a_r^* a_s^*]_+ = 0$$

$$r,s = 1 \ldots M, \quad M = 6(2j+1)$$

The operators a_r are uniquely determined by these conditions; they act in a well-known way on the state function $|\psi\rangle$ which might be expressed as a function of the "occupation" numbers $N_1 \ldots N_r \ldots N_M$ which take the values 0 and 1. In our case the index r stands for a single particle state in a given shell with angular momentum j and represents therefore a combination of three indices, namely the color index k, the index of isotopic spin τ and the index for the "magnetic" quantum number m_j. Combining the indices m_j and τ in one index called α, β or γ assuming now 2(2j+1) values, our pairing operator P in a given shell reads explicitly (summation over repeated indices; i,k,l are distinct color indices and ε is totally antisymmetric)

$$(4)\qquad P = gA^{i*}A^i \quad \text{with} \quad A^i = \varepsilon^{ikl} a_{k\alpha} \eta^{\alpha\beta} a_{l\beta}$$

where $\eta^{\alpha\beta}$ is a symmetric matrix, explicitly given by the well-known pairing expression which in this case, however, had to be generalized in order to include the isotopic spin index. (In conventional pairing theory η is a simplectic matrix; the change-over is due to the inclusion of isospin).

Our task consists now in a rigorous diagonalization of the total Hamiltonian of the system which contains the energies of the independent quarks represented by the "unperturbed" Hamiltonian $H^o = \varepsilon a^*_{i\alpha} a_{i\alpha}$ and the pairing operator P taking, at the same time, the fundamental condition of vanishing total color into account.

The essential point of these calculations consists in the application of general representation-theoretical methods: We thus start with a group theoretical characterisation of the eigen-states of the unperturbed Hamiltonian H^o of the independent particles of an open shell. This eigen-state (omitting, as usual, the contributions from the closed shells) containing N quarks may be written with help of our operators a and satisfying, at the same time, the Pauli principle

$$(5) \qquad |\psi\rangle = \prod_{\rho=1}^{N} a^*_{i_\rho \alpha_\rho} |0\rangle$$

where $|0\rangle$ represents the vacuum state. The ensemble of all these states for any given N yields the representation space for an irreducible representation the group SU(M). The elements of the Lie-algebra of this group operate in the following way (the double indices $i\alpha$ and $k\beta$ must be considered as a single index)

$$(6) \qquad \hat{D} = D_{i\alpha,k\beta} a^*_{i\alpha} a_{k\beta} \quad \text{with} \quad D = -D^+$$

This operator $\hat{D}$ satisfies, in view of the anticommutation relations (3) for a and a^* the same commutation relations as the matrices D and therefore yields by this method special representations of SU(M) for each value of N. (This construction constitutes in - more general form - the basis of so-called second quantization in field theory). The irreducibility follows, according to H. Weyl[4)], from theorems about representation through fully antisymmetrised tensor spaces. In other words, the different eigen-states of H^o are characterized by the corresponding irreducible representations of an underlying "maximal" invariance group of SU(M).

These states, however, do not yet satisfy the fundamental condition about vanishing color. In order to achieve this one has to form characteristic linear combinations of the states given by (5) which read as follows:

$$(7)\quad |\psi\rangle = \prod_{\rho=1}^{n} A^*(\alpha_\rho \beta_\rho \gamma_\rho)\,|O\rangle \text{ with}$$

$$A^*(\alpha\beta\gamma) = \varepsilon^{ikl} a^*_{i\alpha} a^*_{k\beta} a^*_{l\gamma}$$

This implies that the number of quarks must now be a multiple of three, i.e. N = 3n. It is easily checked that these states are invariant with respect to unitary rotations in the 3-dimensional color space.

These new "reduced" states again form a representation space of a special unitary group SU(R), but in this case of the lower dimension R, i.e. M = 3R. SU(R) represents a subgroup of SU(M) and the Lie-elements of this group act through the following operator

$$(8) \qquad \hat{d} = d_{\alpha\beta} a^*_{i\alpha} a_{i\beta} \quad \text{with} \quad d = -d^+$$

yielding again, according to Weyl's theorem about tensor representations with so-called maximal symmetrization, irreducible representations which characterize the eigen-states of the free Hamiltonian H^o in connection with the selection to color-free states.

At this stage one now introduces the pairing interaction term P according to (4). In this way the symmetry of the Hamiltonian is again reduced: It may be readily realized that this new (lower) symmetry group is obtained through the following condition: The characteristic metric η within the pairing operator (4) has to be left invariant under the action of SU(R) characterized by the Lie-elements $d_{\alpha\beta}$ in (8). This additional condition (to be satisfied by the matrix d) defines the subgroup SO(R) (i.e., the orthogonal group) of the original unitary group SU(R). As the eigen-states of H will be characterized by irreducible representations of the invariance group we have - as a next step - to determine the representations which will occur: Acting with our new sub-Lie--algebra on our color-free states according (7) one obtains, first of all, reducible representations of SO(R); the main mathematical or group theoretical problem consists of an explicit determination of the various subspaces corresponding to <u>irreducible</u> representations of the subgroup SO(R), i.e., of splitting the reducible representation into irreducible ones. This can actually be carried through by a method already indicated by H. Weyl[4]: Here we give just an example of the construction of one of these subspaces: It consists, in fact, of the maximal contraction of the color-free states given by (7) (they represent again a tensor space) with help of the metric matrix η:

$$(9) \quad |\psi\rangle = \prod_{\rho=1}^{n} B^*(\alpha_\rho)|0\rangle \quad \text{with} \quad B^*(\alpha) = \varepsilon^{ikl} a^*_{i\alpha} a^*_{k\beta} \eta^{\beta\gamma} a^*_{l\gamma}$$

The further steps containing smaller numbers of contractions and additional reduction were also given by H. Weyl[4]. At this stage it is clear that the eigen--spaces of the full Hamiltonian consist of one or several subspaces corresponding to our irreducible representation of SO(R). It is, however, interesting to note that in our case the various eigen-spaces of H correspond practically always - apart from one exception - to one single irreducible representation of SO(R). This fact is proved by showing that the interaction operator P according (4) represents essentially (apart from trivial terms) the Casimir operator of the group SO(R). This also enables us, at the same time, to determine explicitly the eigen-values of H according to general group theoretical formulas[5]. (The eigen-value of the Casimir operator may be expressed directly by the characters or by the weights of the corresponding irreducible representations).

As a practical application of these group theoretical methods we now consider the explicit structure of the lowest eigen-state of H. It turns out that this state is just the one obtained by the construction given above (i.e., by maximal contraction, cp. formula (9)).

This expression allows a most natural physical interpretation: The ground state of the system is given by n independent 3-quark states with the following properties: Two quarks out of three are paired, i.e., coupled to spin and isospin zero (this coupling is, in fact, expressed by the contraction with the matrix η) whereas the 3rd quark exhibits the characteristic quantum numbers m_j and τ (abbreviated by α) of a nucleon within the conventional theory. One might thus say that the characteristic properties of the original (conventional) shell structure is reproduced: We have just n independent nucleons within our shell. The essential difference consists, however, in the fact that the 3-quark system representing a nucleon within the nucleus is strongly deformed with respect to the quark state of a free nucleon in accordance with the experimental effect[7)] mentioned in section II. (It appears, however, that this deformation is for the moment by far too strong; it will, on the other hand, be reduced if the next steps following this approximation scheme were carried through; they will,at the same time simulate the direct interaction between the nucleons which plays an important role in the conventional scheme). The higher eigen-states of H which follow from our method have also a direct physical interpretation: They represent the inner excitation of nucleons embedded into matter (cp. the remark in sect. II). In conclusion one might say that the characteristic shell structure effect, which appeared at first sight to be typical for the conventional theory, is obtained even in a more natural (group theoretical) way on the basis of the fundamental quark picture[6)].

IV. CONCLUSIONS

We have thus shown that some characteristic nuclear properties - in our case those of the experimentally well-observed shell structure - may be obtained directly from a gauge theoretical viewpoint without using (to some extent questionable) concepts like nuclear force, boson exchange, etc. which so far form the basis of the conventional picture. This fact may strongly suggest that nuclear theory (even in more general cases) could be founded from the outset on the more fundamental principles which form, at the same time, the basis of particle physics; for this reason there should be no distinction between the two domains whatsoever, and the nucleus might just be visualized as a heavy hadron. It must be admitted, however, that this work contains rather rough approximations and simplifying assumptions which are, for the time being, far from being proved: I think about the introduction of special boundary conditions in order to express in a phenomenological way the experimentally well--established confinement, the replacement of the direct interaction between quarks by a pairing force, etc. On the other hand, very similar and partly the same assumptions had already to be made in the quark theoretical treatments of single hadrons. One might thus say that the (approximate) validity of these assumptions is, in the framework of our theory, tested in more general structures, i.e. cases for which by far more experimental data are available.

On the other hand, this test stresses again the necessity of a deeper understanding of various facts in particle physics: The fundamental confinement problem, which is directly related to the basic questions about the quantum chromodynamical vacuum state constitutes the most important example. In this connection the constancy (perhaps universality) of the vacuum pressure B over a wider range of hadronic structures (compare sect. II; it was used in nuclear structure in the same way as in hadron theory) constitutes an interesting hint. At the same time, it might be mentioned that the experimental determination of the (ideally) sharp nuclear shell closures (mentioned in sect. II) appears, by now, directly related to the existence of (ideal) fermions (i.e., the quarks) within nuclear structure, whereas the clustering into three quarks (formula (9), sect. III), which represents the "nucleonic structure" of the nucleus, is a direct consequence of the fundamental (unproved) hypothesis about vanishing total color and the existence of three colors for the single quark; this last fact, in turn, constitutes (as discussed in sect. I) the very basis of gauge theory.

From a more general viewpoint one might thus say that our approximation scheme relates experimental nuclear properties - the nucleus is perhaps the best measured system there is in physics - directly to the fundamental principles of gauge theory. On the other hand, this work shows also that we are only at the very beginning of a deeper understanding of nuclear structure - the nucleus appears to be the most difficult case in this respect - and that there is still a long way to go. For the time being it was, however, appealing to see that rather far-reaching group theoretical methods (as described in extremely abbreviated form in sect. III), developed to a large extent by H. Weyl and explained in his famous book about groups and quantum mechanics, played a decisive role in this research. This part of the work is entirely the merit of Dr. H.R. Petry (Bonn) who will publish it in detail in the near future.

REFERENCES

1) A. Chodos et al., Phys. Rev. D9 3471 (1974)
   T. de Grand et al., Phys. Rev. D12 2060 (1975)
2) Compare for example: A. de Shalit and H. Feshbach Theoretical Nuclear Physics, J. Wiley and Sons, Inc. (1974)
3) M. Beiner, K. Bleuler, R. de Tourreil, Nuov. Cim. 52B 45
4) H. Weyl, Group Theory and Quantum Mechanics, orig. edition: Hirzel Leipzig 1930, compare chapter V
5) N. Jacobson, Lie-algebras, J. Wiley and Sons, New York (1965)
6) K. Bleuler, H. Hofestädt, S. Merk and H.R. Petry, to be published in June 1983 in Zeitschrift für Naturforschung
   K. Bleuler, Proc. 3rd Intern. Conf. on Nuclear Reaction Mechanisms, Varenna 1982, Ricerca Scientifica, Milano 1982
7) J.J. Aubert et al., Cern Report No CERN-EP/83
   A. Bodek et al., Phys. Rev. Lett. 50 (1983) 1431

THEORIES DES JAUGES GRADUEES

Richard Kerner

Département de Mécanique, Université P. et M. Curie,
4 Place Jussieu, 75005 Paris, FRANCE.

Abstract. The mathematical background for a graded extension of gauge theories is investigated. After discussing the general properties of graded Lie algebras and graded Lie groups, the graded fiber bundle is constructed. Its basis manifold is the product of the Minkowskian space-time with the Grassmann algebra spanned by anticommuting Lorentz spinors; the vertical subspaces tangent to the fibers are isomorphic with the graded extension of the SU(N) Lie algebra. The connection and curvature are defined then on this bundle; the two different gradings are either independent of each other, or may be unified in one common grading, which is equivalent to the choice of the spin-statistics dependence.
The Yang-Mills lagrangian is investigated in the simplified case. The conformal symmetry breaking is discussed, as well as other physical consequences of the model.

1. INTRODUCTION

Les théories que nous présentons ainsi sont, en quelque sorte, le croisement entre les théories multidimensionnelles (dont la première, à 5 dimensions, a été développée dans les travaux de Kaluza, Klein, Einstein, Jordan et Thiry(1)) et les théories connues sous le nom des supersymétries, introduites et développées dans les années 1970 par Akulov, Volkov, Wess, Zumino, Salam, Strathdee, Ferrara et autres auteurs (2). Le langage mathématique qui fournit le cadre idéal pour les théories multidimensionelles (dites aussi théories de jauge ou théories des champs de Yang et Mills) est celui des espaces fibrés et des connexions; il a été développé dans les travaux de Lichnerowicz, Chern, Koszul, Nomizu et Kobayashi (3), appliqué en physique théorique par Mme Kerbrat, Trautman, Cho

S. Sternberg (ed.), Differential Geometric Methods in Mathematical Physics, 23–43.

et d'autres auteurs (4). En ce qui concerne les supersymétries, le langage mathématique approprié est celui des algèbres de Lie graduées (cf. Wolf, Corwin, Ne'eman, Sternberg, Rittenberg etc. (5)).
Les algèbres de Lie graduées peuvent être introduites de manière axiomatique; ici nous montrerons leur réalisation particulière en tant que transformations infinitésimales des variétés graduées, pour généraliser ensuite.
Après cette introduction mathématique nous construirons les fibrés gradués ainsi que les connexions dans ces fibrés. Il suffira d'introduire l'analogue du principe variationnel d'Einstein et Hilbert pour pouvoir déduire les équations qui décrivent les intéractions des champs de toutes les espèces, scalaires, vectoriels ... tous réunis dans une seule connexion sur le fibré gradué.
Les problèmes de corrélation entre spin et statistique, de positivité d'énergie, ainsi que de choix particuliers de jauge correspondant à la brisure de symétrie conforme seront discutés sur les exemples les plus simples possibles.

2. ALGEBRE DE LIE GRADUEE ASSOCIEE A UN GROUPE DE LIE COMPACT ET SEMI-SIMPLE.

Soit G un groupe de Lie compact et semi-simple de dimension N ; soit $\mathcal{A}_G$ son algèbre de Lie. Dans un système de coordonnées donné $\mathcal{A}_G$ est engendrée par N générateurs L_a $(a,b,\ldots = 1,2,\ldots,N)$ satisfaisant les relations de commutation

$$[L_a, L_b] = C^d_{ab}\, L_d \tag{1}$$

où $C^d_{ab} = - C^d_{ba}$ sont les constantes de structure du groupe G. Pour tout vecteur $X \in \mathcal{A}_G$, $X = X^a L_a$, où X^a sont les composantes de X dans le repère $\{L_a\}$. Nous avons

$$[X,Y] = - [Y,X] \tag{2}$$

et

$$[[X,Y],Z] + [[Y,Z],X] + [[Z,X],Y] = 0 \tag{3}$$

(identité de Jacobi). En coordonnées locales cela se traduit par

$$C^c_{ab}C^b_{df} + C^c_{db}C^b_{fa} + C^c_{fb}C^b_{ad} = 0 \tag{3a}$$

Introduisons l'application linéaire ad :

$$\mathrm{ad} : \mathcal{A}_G \to L(\mathcal{A}_G, \mathcal{A}_G) \tag{4}$$

par

$$\mathrm{ad}(X)Y = [X,Y] \tag{5}$$

C'est une représentation de $\mathcal{A}_G$ dans les matrices $N \times N$ opérant sur l'espace linéaire $\mathcal{A}_G$:

$$(\mathrm{ad}(X)\mathrm{ad}(Y) - \mathrm{ad}(Y)\mathrm{ad}(X))Z = \mathrm{ad}([X,Y])Z \tag{6}$$

pour tout $X,Y,Z \in \mathcal{A}_G$. (C'est l'identité de Jacobi (3) écrite de manière différente). En coordonnées locales la matrice ad(X) est donnée par

$$\{\mathrm{ad}(X)\}^b{}_d = X^a C^b_{ad} \tag{7}$$

Posons

$$g_G(X,Y) = -\,\mathrm{Trace}(\mathrm{ad}(X)\mathrm{ad}(Y)) = g_G(Y,X) \tag{8}$$

g_G est la forme métrique de Cartan-Killing; pour les groupes compacts et semi-simples elle est non-singulière et définie positive. Soit E un espace vectoriel de dimension s; soit $\tau : \mathcal{A}_G \to L(E,E)$ une représentation fidèle (mais pas forcément irréductible) de $\mathcal{A}_G$:

$$(\tau(X)\tau(Y) - \tau(Y)\tau(X))\, u = \tau([X,Y])u \tag{9}$$

pour tout $X,Y \in \mathcal{A}_G$, $u \in E$.
Nous introduisons la graduation Z_2 en postulant l'anticommutativité des éléments de E :

$$uv = -\,vu \qquad\qquad u,v \in E \tag{10}$$

ainsi que la notion de parité Grassmanienne : on posera $\pi(X) = 0$, $\pi(u) = 1$, etc ...
La réunion $\mathcal{A}_G \oplus E$ possède déjà la structure de produit semi-direct de $\mathcal{A}_G$ avec l'algèbre abélienne E, noté $\mathcal{A}_G \,\square\, E$:

$$\begin{aligned} &(X,Y \in \mathcal{A}_G) \to [X,Y] = -\,[Y,X] \in \mathcal{A}_G \\ &(X \in \mathcal{A}_G,\ u \in E) \to \tau(X)u \in E \\ &(u,v \in E) \to u + v \in E \end{aligned} \tag{11}$$

Remarquons en passant que

$$\pi([X,Y]) = 0 \quad , \quad \pi(\tau(X)u) = 1 \quad . \tag{12}$$

Nous formerons une algèbre graduée si nous introduisons une application bilinéaire symétrique

$$\rho : E \times E \to \mathcal{A}_G. \tag{13}$$

vérifiant

$$\rho(u,v) = \rho(v,u) \in \mathcal{A}_G \tag{14}$$

et

$$[X,\rho(u,v)] = \rho(\tau(X)u,v) + \rho(u,\tau(X)v) \tag{15}$$

Cette dernière formule s'interprête comme dérivation de $\rho(u,v)$ à travers ses arguments (formule de Leibniz). En récapitulant, écrivons encore une fois les formules (3) ou (6), (9) et (15)

$$\begin{aligned} &ad(X)ad(Y)Z - ad(Y)ad(X)Z - ad([X,Y])Z = 0 \\ &\tau(X)\tau(Y)u - \tau(Y)\tau(X)u - \tau([X,Y])u = 0 \\ &\rho(\tau(X)u,v) + \rho(u,\tau(X)v) - [X,\rho(u,v)] = 0 \end{aligned} \tag{16}$$

Pour généraliser l'identité de Jacobi sur $_G$ E il manque encore une formule reliant entre eux trois éléments de E; la voici :

$$\tau(\rho(u,v)w + \tau(\rho(v,w) + \tau(\rho(w,u))v = 0 \tag{16a}$$

Toutes ces formules peuvent être réduites à une seule si l'on introduit le crochet de Lie gradué en définissant

$$\{P,Q\} = -(-1)^{\pi(P)\pi(Q)}\{Q,P\} \tag{17}$$

(où P,Q symbolisent les éléments appartenant soit à $\mathcal{A}_G$ soit à E) comme suit :

$$\begin{aligned} &\text{Si } P = X, Q = Y, \{P,Q\} = \{X,Y\} = [X,Y] = -\{Y,X\} \\ &\text{Si } P = X, Q = u, \{P,Q\} = \{X,u\} = \tau(X)u = -\{u,X\} \\ &\text{Si } P = u, Q = v \ \{P,Q\} = \{u,v\} = \rho(u,v) = \{v,u\} \end{aligned} \tag{18}$$

Dans ce cas, (16) avec (16a) s'écrivent simplement comme

$$\{\{P,Q\},R\} + \{\{Q,R\},P\} + \{\{R,P\},Q\} = 0 \tag{19}$$

La notion de trace ainsi que de la métrique de Cartan-Killing sont aisément généralisées : nous introduisons

$$\mathcal{a}d(P)R = \{P,R\} \tag{20}$$

et la "super-trace" (ou la trace graduée) telle que

$$\mathrm{Str}(\mathcal{a}d(P)\,\mathcal{a}d(R)) = (-1)^{\pi(P)\pi(R)}\mathrm{Str}(\mathcal{a}d(R)\,\mathcal{a}d(P)) \tag{21}$$

d'où il s'en suit facilement que

$$\mathrm{Str}(\mathcal{A}d(X)\,\mathcal{A}d(Y)) = \lambda\, g_G(X,Y) \tag{22}$$

$$\mathrm{Str}(\mathcal{A}d(u)\,\mathcal{A}d(v)) = -\,\mathrm{Str}(\mathcal{A}d(v)\,\mathcal{A}d(u)) = \mu\varepsilon(u,v) \tag{23}$$

où $\varepsilon(u,v) = -\,\varepsilon(v,u)$ est une forme symplectique sur E, tandis que

$$\mathrm{Str}(\mathcal{A}d(X)\,\mathcal{A}d(u)) = 0 \tag{24}$$

de sorte que les sous-espaces de parité différente, $\mathcal{A}_G$ et E, sont orthogonaux. Les facteurs λ et μ dépendent de la représentation τ choisie.

Nous fixerons la représentation dite G-spinorielle en demandant

$$g_G(X,\ \rho(u,v)) = \varepsilon(\tau(X)u,v) = \varepsilon(\tau(X)v,u) \tag{25}$$

ce qui entraine l'invariance de ε :

$$\mathcal{L}_X\, \varepsilon(u,v) = \varepsilon(\tau(X)u,v) + \varepsilon(u,\tau(X)v) = 0 \tag{26}$$

ainsi que

$$g_G(\rho(u,v),\ \rho(w,z)) = \varepsilon(u,w)\varepsilon(v,z) + \varepsilon(u,z)\varepsilon(v,w) \tag{27}$$

Dans ce cas

$$\lambda = \frac{2s(s+1)}{N} \quad , \qquad \mu = 2(s+1) \tag{28}$$

3. EXEMPLE DE REALISATION : OPERATEURS DIFFERENTIELS GRADUES.

L'algèbre de Lie $_G$ est souvent représentée par les champs vectoriels (dérivations) invariants à gauche, engendrés par l'action de G sur lui-même à droite. Par analogie, une algèbre de Lie graduée $\mathcal{A}_G \oplus E$ sera représentée par les dérivations des fonctions définies au dessus d'une variété graduée. Nous commençons par donner l'exemple de super-symétries introduites par Wess, Zumino et al.:

Soit M_4 l'espace-temps de Minkowski, avec la métrique g_{ij} = diag (-+++), i,j = 0,1,2,3 : l'algèbre de Poincaré est réalisée par les opérateurs différentiels agissant sur les fonctions $C^\infty(M_4)$ de manière suivante :

$$\begin{aligned} P_j &= \partial/\partial x_j = \partial_j \\ \mathcal{J}_{k\ell} &= x_k\partial_\ell - x_\ell\partial_k \end{aligned} \tag{29}$$

qui vérifient :

$$[P_j, P_k] = 0 \qquad [\mathbf{J}_{k\ell}, P_m] = g_{km}P_\ell - g_{\ell m}P_k$$
$$[\mathbf{J}_{k\ell}, \mathbf{J}_{mn}] = g_{km}\mathbf{J}_{\ell n} + g_{\ell n}\mathbf{J}_{km} - g_{\ell m}\mathbf{J}_{kn} - g_{kn}\mathbf{J}_{\ell m} \tag{30}$$

La sous-algèbre des rotations $\{\mathbf{J}_{k\ell}\}$ a une représentation spinorielle engendrée par les matrices

$$\sigma_{k\ell} = \frac{1}{8}(\gamma_k\gamma_\ell - \gamma_\ell\,\gamma_k) \tag{31}$$

où γ_k sont les matrices de Dirac, générateurs de l'algèbre de Clifford associée à la métrique gij :

$$\gamma_i\gamma_j + \gamma_j\gamma_i = 2\,gij \tag{32}$$

Dans la notation habituelle, on introduit les indices α, β et $\dot\alpha$, $\dot\beta$: pour distinguer deux représentations 1/2-spinorielles conjointes; γ_i s'expriment alors par les matrices de Pauli de manière suivante

$$(\gamma^j)^\alpha{}_{\dot\beta} = -\sigma^{j\alpha}{}_{\dot\beta}$$
$$(\gamma^j)^{\dot\alpha}{}_\beta = \sigma^j{}_\beta{}^{\dot\alpha} \tag{33}$$
$$(\gamma^j)^\alpha{}_\beta = 0, \quad (\gamma^j)^{\dot\alpha}{}_{\dot\beta} = 0$$

où $\alpha, \beta = 1,2$; $\dot\alpha, \dot\beta = \dot 1, \dot 2$.
Nous levons et baissons les indices α, $\dot\beta$ à l'aide de la "métrique spinorielle" antisymétrique $\varepsilon_{\alpha\beta}$, $\varepsilon_{\dot\alpha\dot\beta}$, ou $\varepsilon^{\alpha\beta}$ $\varepsilon^{\dot\alpha\dot\beta}$, avec

$$\varepsilon_{12} = -\varepsilon_{21} = 1, \quad \varepsilon_{11} = 0, \quad \varepsilon_{22} = 0, \text{ etc } \ldots \tag{34}$$

Dans cette notation

$$\sigma_{o\alpha\dot\beta} = \begin{pmatrix}1&0\\0&1\end{pmatrix}, \quad \sigma_{1\alpha\dot\beta} = \begin{pmatrix}0&1\\1&0\end{pmatrix}, \quad \sigma_{2\alpha\dot\beta} = \begin{pmatrix}0&-i\\i&0\end{pmatrix}, \quad \sigma_{3\alpha\dot\beta} = \begin{pmatrix}1&0\\0&-1\end{pmatrix} \tag{35}$$

L'espace vectoriel dans lequel agit la représentation (31), sera noté $\{\Theta\}$; il est engendré par les éléments de base $\theta^\alpha, \bar\theta^{\dot\beta}$. Nous postulons les relations d'anticommutativité :

$$\theta^\alpha\theta^\beta + \theta^\beta\theta^\alpha = 0, \quad \theta^\alpha\bar\theta^{\dot\beta} + \bar\theta^{\dot\beta}\theta^\alpha = 0, \quad \bar\theta^{\dot\alpha}\bar\theta^{\dot\beta} + \bar\theta^{\dot\beta}\bar\theta^{\dot\alpha} = 0 \tag{36}$$

La variation par rapport aux variables anticommutatives θ est définie comme suit :

$$\partial_\alpha\theta^\beta = \delta^\beta_\alpha, \qquad \partial_\alpha\bar\theta^{\dot\beta} = 0, \text{ etc } \ldots ;$$
$$\partial_\alpha\theta_\beta = \partial_\alpha(\varepsilon_{\gamma\beta}\theta^\alpha) = \varepsilon_{\alpha\beta} \tag{38}$$

avec la loi de Leibniz "graduée", c'est à dire

$$\partial_\alpha(\theta^\beta\theta^\delta) = \delta^\beta_\alpha\theta^\delta - \delta^\delta_\alpha\theta^\beta \quad , \tag{39}$$

il s'en suit que

$$\partial_\alpha\partial_\beta + \partial_\beta\partial_\alpha = 0 \tag{40}$$

Maintenant que nous pouvons définir les opérateurs différentiels généralisés agissant sur les fonctions sur $M_4 \times \{\theta\}$, qui engendrent l'extension graduée de l'algèbre de Poincaré (appelée souvent la "super-algèbre de Poincaré"). Les voici

$$P_j = \partial_j ; \ \mathcal{J}_{k\ell} = x_k P_\ell - x_\ell P_k + \sigma_{k\ell}{}^{\alpha}{}_{\dot\beta}\,\bar\theta^{\dot\beta}\partial_\alpha + \sigma_{\ell\alpha}{}^{\dot\beta}\,\theta^\alpha\partial_{\dot\beta}$$

$$\mathcal{D}_\alpha = \partial_\alpha + \sigma^j{}_{\alpha\dot\beta}\,\bar\theta^{\dot\beta}\,P_j ,\ \bar{\mathcal{D}}_{\dot\beta} = \partial_{\dot\beta} + \sigma^j{}_{\alpha\dot\beta}\,\theta^\alpha\,P_j \tag{41}$$

Les générateurs "pairs" P_j, $\mathcal{J}_{k\ell}$ vérifient les mêmes relations de commutation que précédemment (formules (30)), c'est pourquoi nous avons gardé les mêmes notations. Les nouvelles relations qui définissent l'extension graduée de l'algèbre de Poincaré sont

$$\begin{aligned}
[\mathcal{J}^{mn}, \mathcal{D}_\alpha] &= \sigma^{mn}{}_\alpha{}^{\dot\beta}\,\bar{\mathcal{D}}_{\dot\beta} \\
[\mathcal{J}^{mn}, \bar{\mathcal{D}}_{\dot\beta}] &= \sigma^{mn\alpha}{}_{\dot\beta}\,\mathcal{D}_\alpha \\
[P_j, \mathcal{D}_\alpha] &= 0,\ [P_j, \bar{\mathcal{D}}_{\dot\beta}] = 0 \\
\{\mathcal{D}_\alpha, \mathcal{D}_\beta\}_+ &= 0,\ \{\bar{\mathcal{D}}_{\dot\alpha}, \bar{\mathcal{D}}_{\dot\beta}\}_+ = 0 \ , \\
\{\mathcal{D}_\alpha, \bar{\mathcal{D}}_{\dot\beta}\}_+ &= 2\,\sigma^j{}_{\alpha\dot\beta}\ P_j
\end{aligned} \tag{42}$$

où $\{A,B\}_+ = AB + BA$.
On généralise aisément le calcul extérieur sur $M_4 \times \{\theta\}$. Nous postulons la différentielle extérieure d telle que dd = 0 ; en plus d modifie la parité Grassmanienne de la forme à laquelle il est appliqué. D'où

$$\begin{aligned}
dx^i \wedge dx^j &= -\ dx^j \wedge dx^i \\
dx^i \wedge d\theta^\alpha &= -\ d\theta^\alpha \wedge dx^i \\
d\theta^\alpha \wedge d\theta^\beta &= d\theta^\beta \wedge d\theta^\alpha
\end{aligned} \tag{43}$$

Il est souvent plus facile d'effectuer les calculs dans une base de 1-formes duale aux dérivations ∂_j, $\mathcal{D}_\alpha$ et $\bar{\mathcal{D}}_{\dot\beta}$, telles que

$$e^{\alpha}(\mathcal{D}_{\beta}) = \delta^{\alpha}_{\beta}\ , \quad \bar{e}^{\dot\beta}(\bar{\mathcal{D}}_{\dot\gamma}) = \delta^{\dot\beta}_{\dot\gamma}\ ,$$
$$e^{j}(\partial_k) = \delta^{j}_{k}\ , \tag{44}$$

tandis que $e^{j}(\mathcal{D}_{\alpha}) = 0$, $e^{\alpha}(\partial_{j}) = 0$, etc.

D'où $e^{\alpha} = d\theta^{\alpha}$, $\bar{e}^{\dot\beta} = d\bar{\theta}^{\dot\beta}$,

$$e^{j} = dx^{j} - \sigma^{j}_{\alpha\dot\beta}\theta^{\alpha}\, d\bar{\theta}^{\dot\beta} - \sigma^{j}_{\alpha\dot\beta}\bar{\theta}^{\dot\beta}\, d\theta^{\alpha} \tag{45}$$

Pour terminer, introduisons l'intégration de ces formes en postulant

$$\int d\theta^{\alpha} = 0, \quad \int \theta_{\beta}\, d\theta^{\alpha} = \delta^{\alpha}{}_{\beta}$$

ou bien (46)

$$\int \theta^{\alpha}\, d\theta^{\beta} = \varepsilon^{\alpha\beta}\ , \quad \text{etc}$$

L'élément de volume de super-espace est donc $d^4x \times d^4\theta$, avec

$$\int \theta^{1}\theta^{2}\bar{\theta}^{\dot 1}\bar{\theta}^{\dot 2}\, d\theta^{1}\, d\theta^{2}\, d\bar{\theta}^{\dot 1}\, d\bar{\theta}^{\dot 2} = 1 \tag{47}$$

La même construction peut être faite quant au point de départ sert un groupe de Lie G, compact et semi-simple, de dimension N. La métrique de Cartan-Killing, qui en coordonnées locales s'écrit

$$g_{ab} = C^{d}_{af}\, C^{f}_{db} \qquad (a,b = 1, 2, \ldots, N) \tag{48}$$

engendre l'algèbre de Clifford correspondante :

$$\gamma_a\gamma_b + \gamma_b\gamma_a = 2g_{ab}\ \mathbb{I}d \tag{49}$$

où $\mathbb{I}d$ désigne la matrice identité dans l'espace de représentation Cet espace dans le cas minimal irréductible a la dimension $s = 2^{[N/2]}$, ($[k]$ entier de k); soient ses vecteurs de base $\{\chi^{B}\}$, B,D = 1,2,...,s. Nous appellerons χ^{B} les G-spineurs et postulerons les relations d'anticommutation

$$\chi^{A}\chi^{B} + \chi^{B}\chi^{A} = 0 \tag{50}$$

Ainsi les χ^{B} engendrent la Grassmanienne $\Lambda\{\chi\}$ dont la dimension est 2^{s}.
Dans la base $\{\chi^{B}\}$ les matrices γ_a ont pour éléments $\gamma_{a}{}^{A}{}_{B}$. Il est facile de vérifier que les matrices

$$\sigma_{ab} = \frac{1}{8}(\gamma_a\gamma_b - \gamma_b\gamma_a) \tag{51}$$

engendrent la représentation spinorielle du groupe SO(N)(qui est le groupe d'isométries de la métrique g_{ab}).
Les matrices

$$\tau_a = \frac{1}{2} C^b_{ad}\, g^{df}\, \sigma_{bf} \tag{52}$$

vérifient

$$\tau_a \tau_b - \tau_b \tau_a = C^d_{ab}\, \tau_d \tag{53}$$

et engendrent la représentation fidèle (mais réductible) de $\mathcal{A}_G$ dans l'espace des spineurs χ^B. Ce sont les matrices de l'application τ définie dans § 2.
La forme symplectique invariante ε a pour composantes

$$\varepsilon^{AB} = - \varepsilon^{BA} \tag{54}$$

Si $u = u_A\, \chi^A$, $v = v_B\, \chi^B$, alors

$$\varepsilon(u,v) = - \varepsilon(v,u) = \varepsilon^{AB}\, u_A\, v_B \tag{55}$$

L'inverse de ε^{AB}, ε_{DB} vérifie

$$\varepsilon^{AB}\, \varepsilon_{DB} = \delta^A_D \tag{56}$$

toutes les définitions et formules du § 2 peuvent maintenant s'écrire en coordonnées locales ; notamment, les matrices ρ sont définies par

$$\rho^a_{BD} = \tau^{aG}{}_D\, \varepsilon_{BG} \tag{57}$$

L'extension graduée de $\mathcal{A}_G$ s'obtient de manière identique que celle de l'algèbre de Poincaré. Nous introduisons les dérivations par rapport aux variables G-spinorielles χ^B :

$$\partial_A \chi^B = \delta^B_A, \quad \partial_A(\chi^B \chi^D) = \delta^B_A\, \chi^D - \delta^D_A\, \chi^B \tag{58}$$

Si les N champs vectoriels invariants à gauche qui forment l'algèbre $\mathcal{A}_G$ s'écrivent en coordonnées locales comme

$$S_a = S^b_a\, \partial_b \tag{59}$$

avec

$$S^f_a\, \partial_f\, S^d_b - S^f_b\, \partial_f\, S^d_a = C^f_{ab}\, S^d_f \tag{60}$$

il faut modifier les S_a et y ajouter s générateurs "impairs" pour obtenir l'extension graduée de $\mathcal{A}_G$. Nous introduisons donc

$$\mathcal{D}_a = S_a + \tau_{a\ D}^{\ B} \chi^D \partial_B$$
$$\mathcal{D}_B = \partial_B + \rho^a_{\ BD} \chi^D S_a \tag{61}$$

pour avoir les relations de commutation et anti-commutation suivantes :

$$[\mathcal{D}_a, \mathcal{D}_b] = C^d_{ab} \mathcal{D}_d$$
$$[\mathcal{D}_a, \mathcal{D}_B] = - \tau_{a\ B}^{\ D} \mathcal{D}_D \tag{62}$$
$$\{\mathcal{D}_A, \mathcal{D}_B\}_+ = 2\ \rho^a_{AB} \mathcal{D}_a$$

En désignant par les indices grecs ϕ,ψ les indices (a,B) réunis, pouvant écrire (62) sous forme plus compacte

$$\mathcal{D}_\phi \mathcal{D}_\psi - (-1)^{\pi(\phi)\pi(\psi)} \mathcal{D}_\psi \mathcal{D}_\phi = C^\Omega_{\phi\psi} \mathcal{D}_\Omega \tag{63}$$

avec $\pi(a) = 0$, $\pi(B) = 1$, et $C^\Omega_{\phi\psi}$ données par (62).

Les matrices $(N + s)\times(N + s)$

$$C_a = \begin{pmatrix} C^b_{ad} & 0 \\ 0 & \tau_{aD}^{\ B} \end{pmatrix} \qquad C_B = \begin{pmatrix} 0 & 2\rho^{bD}_B \\ \tau^G_{Ba} & 0 \end{pmatrix} \tag{64}$$

$$(\tau_{B\ a}^{\ G} = - \tau_{a\ B}^{\ G})$$

sont la réalisation adjointe de l'algèbre $\mathcal{A}_G \square E$ graduée.
Nous avons maintenant tout ce qu'il faut pour introduire les théories de jauge graduées.

4. CHAMPS DE JAUGE AVEC GRADUATION.

Nous commençons par quelques brefs rappels sur la théorie de jauge non-abélienne classique. Soit $P(M_4,G)$ fibré principal au-dessus de M_4 avec le groupe de structure G. L'action de G sur $P(M_4,G)$ s'effectue à droite : $P \times G \ni (p,g) \to pg \in P$, et engendre les orbites isomorphes à G ; les champs vectoriels tangents à ces orbites sont invariants à gauche et s'appellent <u>champs verticaux</u>. Soit π : $P(M_4,G) \to M_4$ la projection canonique. Si $X \in TP$ est vertical , $d\pi(X) = 0$. Nous notons $\rho : \mathcal{A}_G \to TP$ l'isomorphisme entre l'algèbre de Lie de G et les sous-espaces verticaux de TP.
Une <u>connexion</u> dans $P(M_4,G)$ est une 1-forme ω à valeurs dans $\mathcal{A}_G$, invariante à gauche :

$$\mathcal{L}_X\, \omega = - \operatorname{ad}(X)\, \omega \tag{65}$$

pour tout X vertical dans TP ; sous forme intégrale on peut écrire

$$\omega(pg) = ad(g^{-1})\omega(p) \tag{66}$$

Un vecteur $X \in TP$ est dit <u>horizontal</u> si

$$\omega(X) = 0 \tag{67}$$

Tout vecteur de TP peut alors être décomposé de manière invariante en deux parties, horizontale et verticale :

$$X = \text{hor}\, X + \text{ver}\, X = (X - \sigma \circ \omega(X)) + \sigma \circ \omega(X) \tag{68}$$

La <u>courbure</u> de la connexion ω est définie par

$$\Omega(X,Y) = d\omega(\text{hor}\, X\, ,\ \text{hor}\, Y) \tag{69}$$

et vérifie l'identité

$$\Omega(X,Y) = d\omega(X,Y) + \frac{1}{2}[\omega(X),\omega(Y)]_{\mathcal{A}_G} \tag{70}$$

grâce à l'invariance de ω.
L'existence de la connexion ω dans $P(M_4,G)$ permet de définir une métrique canonique sur P à partir de la métrique dans M_4 et de la forme de Cartan-Killing dans G : pour $X,Y \in TP$ on pose

$$g_p(X,Y) = g_{M_4}(d\pi(X),d\pi(Y)) + g_G(\omega(X),\omega(Y)) \tag{71}$$

La courbure scalaire de g_p se décompose en trois termes suivants :

$$R_p = R_{M_4} + L(\Omega) + g_G^2 \tag{72}$$

(0 si M_4 Minkowskien)

où

$$g_G^2 = g_{ab}\, g^{ab} = N = \dim G,$$

et

$$L(\Omega) = -\frac{1}{4}\mathrm{Tr}\,(g^{ij}g^{kl}\Omega_{ik}\Omega_{jl}) = -\frac{1}{4}g^{ij}g^{kl}g_{ab}\Omega^a_{ik}\Omega^b_{jl} \tag{73}$$

Ici g^{ij} sont les composantes de la métrique g_{M_4}, $(i,j = 0,1,2,3)$; g_{ab} sont les composantes de la métrique g_G, et

$$\Omega^a_{ij} = \partial_i\, \omega^a_j - \partial_j \omega^a_i + C^a_{bd}\omega^b_i\omega^d_j \tag{74}$$

sont les composantes de la courbure (appelée aussi le champ de Yang et Mills) ; $L(\Omega)$ est le lagrangien de champ de jauge. Le principe variationnel généralisé au $P(M_4,G)$ s'écrit donc (nous négligeons la constante g_G^2) comme suit :

$$\delta \int_{P(M_4,G)} \sqrt{|g_p|} \quad L(\Omega)\ d^4x\ dG = 0 \tag{75}$$

En fait, la quantité intégrée étant invariante par l'action du groupe, l'intégration par rapport à la mesure de Haar dG fait apparaître une constante multiplicative ; ainsi le problème variationnel sur M_4 (rappelons aussi $|g_P| = |\det g_p| = |\det g_{M_4}| \times |\det g_G|$), ce qui donne les équations de Yang et Mills bien connues :

$$\partial^i \Omega^a_{\cdot\, ij} + C^a_{bd}\ \omega^{bi}\ \Omega^d_{\cdot\, ij} = 0 \tag{76}$$

La généralisation formelle est immédiate, sauf l'existence globale du fibré gradué. Nous prenons le cas le plus simple, où le fibré gradué est conçu comme le produit Cartésien de $M_4 \times \{\theta\}$ (la base) avec $G \,\square\, \Lambda\, \{\chi\}$ ("groupe gradué"). La connexion graduée est donc une 1-forme sur ce produit, à valeurs dans l'algèbre graduée $\mathcal{A}_G \,\square\, \{\chi\}$.

En introduisant les indices généralisés K,L,... pour désigner (α,β,j) (indices "horizontaux" du groupe gradué), nous pouvons écrire la 1-forme de connexion sur le fibré gradué comme

$$\begin{aligned} A = A_K e^K = A^\phi_K e^K \mathcal{D}_\phi = (A^a_{\cdot j}\ e^j + A^a{}_\alpha e^\alpha + A^a_\beta\ e^\beta)\, \mathcal{D}_a \,+ \\ + (A^B_{\cdot j}\ e^j + A^B_\alpha\ e^\alpha + A^B_\beta\ e^\beta)\, \mathcal{D}_B \end{aligned} \tag{78}$$

les coefficients A^ϕ_K étant des fonctions de $\chi^B, \theta^\alpha, \theta^\beta, x^j$ et ξ^b (ξ^b étant les paramètres dans G), et en tant que telles peuvent avoir ou pas une parité Grassmanienne déterminée. En plus, il y aura deux parités distinctes, une provenant des puissances de variables χ^B, l'autre des puissances de variable θ^α ; nous devons arrêter le choix entre deux situations différentes, celle où les variables χ^B et θ^α commutent entre elles (alors les parités Grassmaniennes correspondantes se multiplient séparément avant de s'ajouter) ou bien χ^B et θ^α anticommutent entre elles (alors les parités Grassmaniennes s'ajoutent avant de se multiplier).

Pour que les généralisations de formule (70), (74) puissent avoir un sens dans l'extension graduée, il faut que les coefficients A^ϕ_K aient les parités bien définies, c'est-à-dire admettent dans leur développement certaines puissances de χ, θ seulement, et non pas toutes les puissances possibles.

Considérons la formule (74) généralisée : nous devons avoir

$$F^\phi_{KL} = \mathcal{D}_L\ A^\phi_L - (-1)^{\pi(K)\pi(L)} \mathcal{D}_L\ A^\phi_K + C^\phi_{\psi\Omega}\ A^\psi_K\ A^\Omega_L \tag{79}$$

Nous avons arrêté les propriétés de symétrie de dA puisque

$dx^i \wedge dx^j = - dx^j \wedge dx^i, dx^i \wedge d\theta^\alpha = - d\theta^\alpha \wedge dx^i, d\theta^\alpha \wedge d\theta^\beta = d\theta^\beta \wedge d\theta^\alpha$.
Il est naturel de vouloir étendre cette symétrie à la différentielle covariante de A donnée par (79). Les propriétés de symétrie des constantes de structure graduées étant bien définies :

$$C^\phi_{\psi\Omega} = - (-1)^{\pi(\psi)\pi(\Omega)} C^\phi_{\Omega\psi} \tag{80}$$

nous pouvons fixer la symétrie exigée pour le terme $C^\phi_{\psi\Omega} A^\psi_K A^\Omega_L$ en définissant bien les relations de commutation (ou anticommutation) des A^ψ_K. Voici dans les choix possibles et leurs réalisations : nous verrons qu'il est nécessaire de fixer une jauge particulière pour les A^Ω_L (en fait, une classe des jauges est fixée déjà si l'on demande la parité déterminée pour les A^Ω_L).

1) Premier cas de figure : $\chi^B \theta^\alpha = \theta^\alpha \chi^B$, les variables spinorielles du groupe G et de la base M_4 commutent entre elles. Dans ce cas

$$A^\psi_K A^\Omega_L = (-1)^{[\pi(\psi)\pi(\Omega) + \pi(K)\pi(L)]} A^\Omega_L A^\psi_K \tag{81}$$

ce qui donne le tableau correspondant :

	A^b_j	A^b_α	A^B_j	A^B_α
A^d_i	0	0	0	0
A^d_β	0	1	0	1
A^D_i	0	0	1	1
A^D_β	0	1	1	0

(0 correspondant à la commutation,
1 correspondant à l'anti-commutation)

Pour assurer la positivité de l'énergie dans le cas stationnaire, il faut que la connexion soit hermitienne, c.à.d.

$$(A^\phi_\alpha) = (A^\phi_\alpha)^+ \tag{82}$$

où

$$(\theta_\alpha)^+ = \bar{\theta}_{\dot\alpha} , \quad (\theta_\alpha \theta_\beta)^+ - \bar{\theta}_{\dot\beta} \bar{\theta}_{\dot\alpha} , \text{ etc.} \tag{83}$$

Les composantes A^ϕ_α peuvent dépendre des χ^B d'une façon bien déterminée seulement, pour assurer l'invariance à gauche par l'action de la super-algèbre $\mathcal{A}_G \,\square\, \{\chi_B\}$.
Sans donner la preuve (qui se fait par le calcul essentiellement) nous donnons l'exemple de réalisation des A^ϕ_K vérifiant (81) et (82). Ainsi, la composante A^b_j pourrait, en général, comprendre les termes suivants :

$$A^{b}_{\dot{j}}(x,\theta) = B^{b}_{\dot{j}}(x) + B^{b}_{\dot{j}\alpha}(x)\theta^{\alpha} + B^{b}_{\dot{j}\dot{\beta}}(x)\bar{\theta}^{\dot{\beta}} + \psi^{b}_{\alpha}(x)\sigma_{\dot{j}}{}^{\alpha}{}_{\dot{\beta}}\bar{\theta}^{\dot{\beta}} + \psi^{b}_{\dot{\beta}}(x)\sigma_{\dot{j}\alpha}{}^{\dot{\beta}}\theta^{\alpha} + \ldots$$
$$+ \ldots \text{ (jusqu'à la quatrième puissance des } \theta) \tag{84}$$
$$+ \psi^{B}_{\dot{j}}(x)\tau^{a}_{BD}\,\chi^{D} + \ldots \text{ (les puissances des } \chi)$$

ce qui admettrait l'existence de toute sorte de champs ; mais dans ce cas $A^{b}_{\dot{j}}$ n'aurait pas les propriétés requises par (81) et (82). Ces conditions sont si fortes, qu'elles ne laissent la place qu'aux quelques formes seulement :

$$\begin{aligned} A^{b}_{\dot{j}} &= B^{b}_{\dot{j}}(x) \\ A^{b}_{\alpha} &= \phi^{b}(x)\theta_{\alpha} \quad , \quad A^{b}_{\dot{\beta}} = \phi^{b}(x)\bar{\theta}_{\dot{\beta}} \\ A^{B}_{\dot{j}} &= \chi^{B}W_{\dot{j}}(x) \\ A^{B}_{\alpha} &= \chi^{B}\theta_{\alpha}D(x) \quad , \quad A^{B}_{\beta} = \chi^{B}\theta_{\beta}\,D(x) \end{aligned} \tag{85}$$

Nous voyons apparaître le champ de jauge classique $B^{b}_{\dot{j}}$, le multiplet scalaire ϕ^{b} appelé le champ de Higgs, un champ scalaire D(x) et un champ vectoriel $W_{\dot{j}}(x)$. Dans le cas où A^{ϕ}_{K} ne dépendent pas de $\{\chi\}$ explicitement, il ne reste que le champ de Yang et Mills $B^{b}_{\dot{j}}$ et le champ de Higgs. Après avoir calculé les composantes de F^{ϕ}_{KL} et en prenant

$$\mathrm{Str}\,(g^{KM}g^{LN}F_{KL}F_{MN}) \tag{86}$$

nous obtenons l'analogie du lagrangien de la théorie de jauge classique (73). Pour que l'analogie soit totale, il faut intégrer maintenant par rapport au "super volume" $d^4xd^4\theta$, ce qui fait que, grâce à la relation (47), seule comptera la contribution qui dans (86) contient $\theta^1\theta^2\bar{\theta}^1\bar{\theta}^2$. Il est facile de calculer que cette contribution est égale (à un facteur près) à

$$\begin{aligned} L_{(4)} = &- \frac{1}{4}\,\Omega^{a}_{ij}\Omega^{ij}_{a} - \nabla_{i}\phi^{a}\,\nabla^{i}\phi_{a} - \frac{1}{2}\,\partial_{i}D\partial^{i}D \\ &- \frac{1}{4}(\partial_{i}W_{j} - \partial_{j}W_{i})(\partial^{i}W^{j} - \partial^{j}W^{i}) \end{aligned} \tag{87}$$

où

$$\Omega^{a}_{ij} = \partial_{i}B^{a}_{j} - \partial_{j}B^{a}_{i} + C^{a}_{bd}B^{b}_{i}B^{d}_{j} \tag{88a}$$

$$\nabla_{i}\phi^{a} = \partial_{i}\phi^{a} + C^{a}_{bd}B^{b}_{i}\phi^{d} \tag{88b}$$

et correspond au lagrangien classique de la théorie. Toutefois , c'est une théorie bien pauvre, puisqu'elle ne contient que les champs de spin entier (0 ou 1), et ne contient pas les fermions. Ces derniers vont apparaître si nous faisons l'hypothèse alternative à (81), à savoir :

2) Deuxième cas de figure : $\chi^B \theta^\alpha = - \theta^\alpha \chi^B$, les variables spinorielles de deux espèces différentes anticommutent (ce qui mérite le nom de "l'unification"). Dans ce cas

$$A_K^\phi A_L^\psi = (-1)^{(\pi(\phi)+\pi(K))((\pi(\psi)+\pi(L))} A_L^\phi A_K^\psi \tag{89}$$

ce qui donne le tableau suivant :

	$A^b_{\cdot j}$	A^b_α	$A^B_{\cdot j}$	A^B_α
$A^d_{\cdot i}$	0	0	0	0
A^d_β	0	1	1	0
$A^D_{\cdot i}$	0	1	1	0
A^D_β	0	0	0	0

Maintenant les bonnes propriétés de symétrie de l'expression $C_\psi{}^\phi{}_\Omega A^\psi_K A^\Omega_L$ ne peuvent pas être obtenues qu'en imposant une condition de jauge supplémentaire :

$$A^b_\alpha = 0 \quad , \quad A^b_{\dot\beta} = 0 \quad , \quad A^B_{\cdot j} = 0 \tag{90}$$

(uniquement les composantes "paires" restent non-nulles).
Avec la condition d'hermicité ceci donne le développement suivant :

$$\begin{aligned} A^a_{\cdot j} &= B^a_{\cdot j}(x) + \chi^A \tau^a_{AB}(\psi^B_\alpha(x)\sigma_{\cdot j}{}^\alpha{}_{\dot\beta}\bar\theta^{\dot\beta} + \theta_\alpha \sigma_{\cdot j}{}^\alpha{}_{\dot\beta}\bar\psi^{B\dot\beta}) + \\ &\quad \chi^A \tau^a_{AB}(\psi^A_{\cdot j\alpha}\theta^\alpha + \bar\psi^A_{\cdot j\dot\beta}\bar\theta^{\dot\beta}) \\ A^B_\alpha &= \psi^B_\alpha(x) + \psi^B_{\cdot j\alpha}\sigma^j_{\gamma\dot\delta}\theta^\gamma\bar\theta^{\dot\delta} + D(x)\chi^B\theta_\alpha + W_{\cdot j}(x)\sigma^j_{\alpha\dot\beta}\bar\theta^{\dot\beta}\chi^B + \\ &\quad + \phi^a(x)\tau^B_{aD}\chi^D\theta_\alpha + \tau^B_{aD}\chi^D B^a_{\cdot j}(x)\,\sigma^j_{\alpha\dot\beta}\bar\theta^{\dot\beta} \end{aligned} \tag{91}$$

et $A^B_{\dot\alpha} = (A^B_\alpha)^+$

Rappelons que cette jauge n'a de sens que dans le système non-holonome covariant e^j, e^α, $\bar e^\beta$ défini par (45).

Nous voyons apparaître les champs de spin non-entier à côté des champs de spin entier ; leur multiplicité (c.a.d. la représentation du groupe G à laquelle ils appartiennent) est, quant à elle, déterminée par les conditions de commutation (89) : les champs bosoniques se transforment soit suivant la représentation triviale $\mathcal{A}_G \to \mathbb{I}d$ (champ scalaire D(x), champ vectoriel $W_j(x)$, soit suivant la représentation adjointe (champ de Yang-Mills B^a_j, champ de Higgs ϕ^a) ; tandis que les champs fermioniques se transforment suivant la représentation spinorielle du groupe G (les fermions ψ^B_α de Dirac et les vecteurs-spineurs $\psi^B_{j\alpha}$).

Dans le lagrangien défini par (86) nous pouvons retenir soit la partie contenant $\theta^1\theta^2\bar{\theta}^1\bar{\theta}^2$ et ne contenant pas les G-spineurs χ ; soit la partie contenant $\varepsilon_{AB}\chi^A\chi^B\theta^1\theta^2\bar{\theta}^1\bar{\theta}^2$ qui généralise, en quelque sorte, le fait que l'on intègre aussi selon les directions correspondant au groupe gradué. C'est cette partie là qui donne le lagrangien classique (87) plus les termes correspondant au lagrangien de Dirac :

$$\varepsilon_{AB}(\psi^A\gamma^j\nabla_j\bar{\psi}^B - (\nabla_j\psi^A\gamma^j)\bar{\psi}^B) \tag{92}$$

où

$$\nabla_j\psi^A = \partial_j\psi^A + \tau^A_{bB}B^b_j\psi^B \tag{93}$$

(nous n'avons effectué les calculs que pour une version simplifiée sans $\psi^A_{j\alpha}$).

Tous les champs dans ce lagrangien sont de masse nulle, ce qui n'est pas étonnant vu que nos expressions sont invariantes par rapport à l'action du groupe conforme et qu'aucun paramètre naturel de dimension cm^{-1} n'a été introduit.

En résumant, nous pouvons regrouper nos champs dans le tableau suivant :

G-spin \ Spin dans espace-temps	0	1/2	1	3/2	2
0	D(x)	-	$W_j(x)$	-	$h_{ij}(x)$
1/2	-	$\psi^A_\alpha(x)$	-	$\psi^A_{j\alpha}(x)$	-
1	$\phi^b(x)$	-	$B^b_j(x)$	-	$h^b_{ij}(x)$

Nous nous sommes permis de prolonger le tableau vers le spin spatio-temporel 2. Logiquement, si l'on ajoutait le champ de spin 2 (tenseur symétrique h_{ij} de trace nulle), il y a aussi la place pour le multiplet adjoint h^b_{ij}. Pour l'instant nous ignorons quel genre de particules élémentaires cela pourrait représenter (surtout que nous ne sommes pas certains quel est le groupe G).

5. JAUGES BRISANT LA SYMETRIE CONFORME. APPARITION DE MASSES.

Jusqu'ici nous n'avons pas parlé de dimension de nos champs ; cela n'a pas été nécessaire puisque toutes nos expressions étaient invariantes par rapport à l'action du groupe conforme, notamment les dilatations ; aucune constante élémentaire dimensionnelle n'entrait dans les développements (91). Pourtant, l'introduction d'une telle constante s'avère tout à fait naturelle.

Commençons par arrêter la dimension des variables spatio-temporelles : $x^j = (ct,x,y,z)$. Evidemment, dim $|x^j|$ = cm ; symboliquement dim $[\partial/\partial x^j] = \mathrm{cm}^{-1}$. Le tenseur de champ de jauge, F^a_{ij}, a dans ces unités la dimension cm^{-2} traditionnellement. Puisque

$$F^a_{ij} = \partial_i A^a_j - \partial_j A^a_i + C^a_{bd} A^b_i A^d_j \quad ,$$

nous devons avoir dim $A^a_j = \mathrm{cm}^{-1}$ et supposer les constantes C^a_{bd} sans dimension. (On utilise parfois un autre choix, où A^a_i sont sans dimension, les constantes C^a_{bd} sont multipliées par une constante dimensionnelle e, dim$[e] = \mathrm{cm}^{-1}$, et le F^a_{ij} résultant est multiplié à son tour par e).

Nous supposons que les composantes $\sigma^j_{\alpha\beta}$ sont des nombres sans dimension. Dans ce cas il est clair que, puisque

$$\dim\,[\theta^\alpha] = \frac{1}{\dim\,|\partial_\alpha|} \tag{94}$$

et

$$\begin{aligned} \dim\,[\mathscr{D}_\alpha] &= \dim\,[\partial_\alpha + \sigma^j_{\alpha\dot\beta}\bar\theta^{\dot\beta}\,\partial_j] \\ &= (\dim[\theta]^{-1} + (\dim[\theta])\,\mathrm{cm}^{-1} = (\dim[\theta])^{-1} \end{aligned} \tag{95}$$

il faut que dim $[\theta] = \mathrm{cm}^{1/2}$.

Les spineurs sont, en quelque sorte, des "racines carrées" de vecteurs ; ainsi

$$\dim\,[\sigma^j_{\alpha\dot\beta}\theta^\alpha\bar\theta^{\dot\beta}] = 2\,\dim\,[\theta] = \mathrm{cm} \tag{97}$$

D'autre part, la dimension de la densité lagrangienne est égale à cm^{-4}, de sorte que l'intégrale variationnelle soit sans dimension : dim $|Ld^4x| = 0$, d'où encore dim $[F^a_{ij}.F^{ij}_a] = \mathrm{cm}^{-4}$. Quand il y a la contribution de champ de Higgs, elle est égale à $-(1/2)D_j\phi^a D^j\phi_a$; puisque Dim $[D_j] = \mathrm{cm}^{-1}$, dim $|\phi^a| = \mathrm{cm}^{-1}$ aussi. Donc dans la formule (85), dim $[A^a_\alpha]$ = dim$[\phi^a\,\theta^\alpha] = \mathrm{cm}^{-1}.\mathrm{cm}^{1/2} = \mathrm{cm}^{-1/2}$, et dim $|F^a_{\alpha j}| = \mathrm{cm}^{-3/2}$, dim$[F^a_{\alpha\beta}] = \mathrm{cm}^{-1}$. Pour pouvoir additionner les contributions $F^a_{ij}.F^{ij}_a$, $F^a_{i\alpha}\,F^{i\alpha}_a F^{\alpha\beta}_a$ il fallait introduire un facteur dimensionnel ℓ, dim $|\ell|$ = cm, comme suit :

$$F^{a}_{ij}{}^{\cdot}F^{ij}_{a} + \frac{1}{\ell} F^{a}_{i\alpha} F^{i\alpha}_{a} + \frac{1}{\ell^2} F^{a}_{\alpha\beta} F^{\alpha\beta}_{a} \tag{98}$$

Dans le cas du développement (85) le terme contenant $\theta^1\theta^2\bar{\theta}^{\dot{1}}\bar{\theta}^{\dot{2}}$ (le seul qui nous intéressait) était homogène en $1/\ell^2$, qui n'apparaissait plus dans les équations produites par le principe variationnel.

Il n'en n'est pas de même si nous introduisons les termes contenant ℓ explicitement dans le développement de la connexion graduée, brisant par là même la symétrie conforme. Considérons l'exemple le plus simple, qui dérive de la théorie sans fermions, présentée par les formules (81) et (85). Tout en respectant l'hermiticité et les conditions sur la parité Grassmanienne, nous pouvons choisir :

$$A^{b}_{\dot{j}} = B^{b}_{\dot{j}}(x) + \frac{f}{\ell} \phi^{b} \sigma_{j\alpha\dot{\beta}} \theta^{\alpha}\bar{\theta}^{\dot{\beta}}$$

$$A^{b}_{\alpha} = \phi^{b}(x)\theta_{\alpha} + \frac{f}{\ell} \phi^{b} \theta_{\alpha}\bar{\theta}_{\dot{\beta}}\bar{\theta}^{\dot{\beta}} \ , \quad A^{b}_{\dot{\alpha}} = (A^{b}_{\alpha})^{+} \tag{99}$$

$$A^{B}_{\dot{j}} = 0 \ , \quad A^{B}_{\alpha} = 0 \ , \quad A^{B}_{\dot{\beta}} = 0$$

f est une constante sans dimension, ℓ l'échelle de longueur brisant la symétrie conforme. Le nouveau lagrangien contient un terme de masse pour le champ ϕ^b :

$$L = -\frac{1}{4} \Omega^{a}_{ij} \Omega^{ij}_{a} - (\frac{3f^2}{8} + \frac{1}{2})\nabla_i\phi^a\nabla^i\phi_a - \frac{4}{\ell^2} \phi^a\phi_a \tag{100}$$

Pour interpréter ces nouveaux termes comme avant, il faut renormaliser le champ ϕ^a en introduisant $\tilde{\phi}^a$ de telle sorte que :

$$(\frac{3f^2}{8} + \frac{1}{2}) \nabla_i\phi^a\nabla^i\phi_a = \frac{1}{2} \nabla_i\tilde{\phi}^a\nabla^i\tilde{\phi}_a \tag{101}$$

Le choix le plus simple est offert par f = 2, $\tilde{\phi}^a = 2\phi^a$; dans ce cas nous identifions :

$$L = -\frac{1}{4} \Omega^{a}_{ij}{}^{\cdot}\Omega^{ij}_{a} - \frac{1}{2} \nabla_i\tilde{\phi}^a\nabla_i\tilde{\phi}_a - \frac{1}{\ell^2} \tilde{\phi}^a\tilde{\phi}_a \tag{102}$$

ce qui donne la masse du champ de Higgs à $m_\phi = \sqrt{2}/\ell$. Le champ $B^{a}_{\dot{j}}$ est resté de masse nulle.

La même technique s'applique au cas plus réaliste, présenté par le développement (91). Dans ce cas, il faut remplacer :

$$\begin{aligned}
\phi^a &\to \phi^a + \frac{f}{\ell}\,\phi^a(\theta_\alpha\theta^\alpha + \bar\theta_{\dot\beta}\bar\theta^{\dot\beta}) \\
W_J &\to W_j + \frac{\omega}{\ell}\,W_j(\theta_\alpha\theta^\alpha + \bar\theta_{\dot\beta}\bar\theta^{\dot\beta}) + \frac{d}{\ell}\,D\,\sigma_{j\alpha\dot\beta}\theta^\alpha\bar\theta^{\dot\beta} \\
D &\to D + \frac{d}{\ell}\,D(\theta_\alpha\theta^\alpha + \bar\theta_{\dot\beta}\bar\theta^{\dot\beta}) + \frac{\omega}{\ell}\,W_j\,\sigma^j_{\alpha\dot\beta}\theta^\alpha\bar\theta^{\dot\beta} \qquad (103) \\
\psi^B_\alpha &\to \psi^B_\alpha + \frac{p}{\ell}\,\psi^B_\alpha(\theta_\gamma\theta^\gamma + \bar\theta_{\dot\gamma}\bar\theta^{\dot\gamma})\ ,
\end{aligned}$$

et ainsi de suite. Notre théorie devient alors plus symétrique, elle mélange plus les champs entre eux, mais la symétrie conforme est brisée et les masses apparaissent (sauf pour les champs B^a_j et W_j qui restent de masse nulle). Les détails de calculs seront publiés ultérieurement (6), (7). Nous montrons ainsi, pour donner un exemple, le lagrangien d'une théorie simplifiée (sans champ scalaire, vectoriel, ni de spin 3/2) :

$$\begin{aligned}
A^a_j &= B^a_j(x) + \frac{1}{\ell}\,\phi^a(x)\sigma_{j\alpha\dot\beta}\,\theta^\alpha\bar\theta^{\dot\beta} \\
A^a_\alpha &= \phi^a(x)\theta_\alpha + \sigma^j_{\alpha\dot\beta}\bar\theta^{\dot\beta}\,A^a_\alpha\ , \quad A^a_{\dot\alpha} = (A^a)^+ \\
A^B_j &= 0 \qquad (104) \\
A^B_\alpha &= \psi^B_\alpha(x) + \frac{\lambda}{\ell}\,(\psi^B_\gamma\theta^\gamma + \bar\psi^B_{\dot\gamma}\bar\theta^{\dot\gamma})\,\theta_\alpha + \frac{\gamma}{\ell^2}\,\psi^B_\alpha\,\theta^1\theta^2\bar\theta^{\dot 1}\bar\theta^{\dot 2} \\
A^B_{\dot\alpha} &= (A^B_\alpha)^+
\end{aligned}$$

Après les calculs, on obtient

$$\begin{aligned}
L = &-\frac{1}{4}\,\Omega^a_{ij}\,\Omega^{ij}_a - \frac{1}{2}\,\nabla_j\tilde\phi^a\nabla^j\tilde\phi_a - \frac{1}{8\ell^2}\,\tilde\phi^a\,\tilde\phi_a - \frac{2s}{N}\Big[\frac{\lambda^2(\lambda+2)}{4l^3}\,\psi\psi + \\
&+ \frac{\lambda(3\lambda+2)}{8\ell^2}\,(\psi\gamma^j\nabla_j\bar\psi - (\gamma^j\nabla_j\psi)\bar\psi) + \frac{\lambda}{8\ell}\,\nabla_j\tilde\phi^a C_{aBD}\psi^B\gamma^j\bar\psi^D\Big] - \qquad (105) \\
&- \frac{\lambda(\lambda+2)}{16\,\ell^2}\,\tilde\phi^a C_{aBD}\psi^B\psi^D - \frac{\lambda}{2\ell^2}\,C^a{}_{BD}C_{aEG}\psi^B\psi^D\bar\psi^E\bar\psi^F
\end{aligned}$$

où

$$\psi\,\bar\psi = \varepsilon_{AB}(\varepsilon^{\alpha\beta}\psi^A_\alpha\psi^B_\beta + \varepsilon^{\dot\alpha\dot\beta}\bar\psi^A_{\dot\alpha}\bar\psi^B_{\dot\beta}\ ),$$

$$\nabla_j\psi^B = \partial_j\psi^B + \tau_a{}^B{}_D\,B^a_j\psi^D\ , \quad \text{etc}$$

Nous pouvons observer les termes de masse pour les changes ϕ^a et ψ^B_α ; leur rapport est égal à :

$$\frac{m_\phi}{m_\psi} = \frac{2N}{2s(\lambda^2(\lambda+2))} = \frac{N}{3}\,\frac{1}{\lambda^2(\lambda+2)} \tag{106}$$

avec $(\lambda(3\lambda+2)/8).(2s/N) = 1/2$ pour renormaliser le champ ψ de telle sorte que l'on ait :

$$\frac{1}{2}\left[\psi\,\gamma\,\nabla_j\bar{\psi} - (\gamma^j\,\nabla_j\psi)\,\bar{\psi}\right] \tag{107}$$

dans le terme correspondant au lagrangien de Dirac. Le rapport m_ϕ/m_ψ dépend donc de la dimension du groupe G choisi, N (rappelons que s = 2 à la puissance $[N/2]$.

Dans le cas le plus simple où G = SU(2), N = 3, s = 2, m_ϕ/m_ψ = 0.0647 ; dans le cas où G = SU(3), N = 8, s = 16, m_ϕ/m_ψ = 0.1011. Pour comparer, dans la théorie des forces nucléaires, où G = SU(2) et où l'on identifie ϕ^a avec le triplet de mésons $\pi(\pi^+,\pi^\circ,\pi^-)$ et le couple ψ^1,ψ^2 avec les 2 nucléons (p et n, proton et neutron), $m_\pi/m_p \simeq 0.147$. Nous pouvons nous consoler en nous disant que les ordres de grandeur ne sont pas trop mauvais, surtout qu'en réalité les masses observées doivent être corrigées par les effets des intéractions diverses.

Pour conclure, quelques mots sur les théories unifiées et les constantes dimensionnelles. Pour l'instant, nous avons dû introduire une constante dimensionnelle ℓ pour avoir une commune mesure entre les dimensions spatio-temporelles x^j et les spineurs θ^α. Dans l'esprit "unifié" nous devons admettre que toutes les dimensions dans la variété fibrée P(M,G) sont de même nature, c.a.d. ont la même dimension |cm|. Ainsi, nous devons introduire une nouvelle constante dimensionnelle, k, pour donner la dimension |cm| aux variables dans le groupe G. Tertio, une fois que les variables ξ^b du groupe G se transforment en $k\xi^b$ et ont cette dimension acquise, il faut donner la dimension $cm^{1/2}$ aux G-spineurs χ^B, de sorte que $\varepsilon_{AB}\chi^A\chi^B$ ait une "longueur" g. Nous voyons donc apparaître trois constantes dimensionnelles, ℓ , k et g, ou bien trois nombres fondamentaux (ℓ/k), (ℓ/g), (k/g), de manière assez naturelle. Reste à expliquer les valeurs observées que l'on aimerait à faire correspondre à des interactions gravitationnelles, electro-faibles et fortes.

REFERENCES

(1) Kaluza, Th., 1921, Berl. Berichte, pp. 966.
Einstein, A., 1927, Sitz. Preuss. Akad. Wiss., 23-25,26-30.
Klein, O., 1926, Z. Phys. 37, p. 895.
Thiry, Y., 1948, C.R. Acad. Sci. Paris, 226, p. 216.

(2) Volkov, D.V., Akulov, V.P., 1973, Phys. Lett. 46B, p. 109.
Wess, J., Zumino, B., 1974, Nucl. Phys. 70B, p. .
Salam, A., 1974, Nucl. Phys., B76, p. 477.
Arnowitt, R., Nath, P., Zumino, B., 1975, Phys. Lett. B56 p. 81.
Ferrara, S., Zumino, B., 1974, Nucl. Phys., B79, p. 413.
Iliopoulos, J., Zumino, B., 1974, Nucl. Phys., B 76, p. 310.
Gawedzki, K., 1977, Ann. Inst. H. Poincaré, 27 (4), p. 335.

(3) Lichnerowicz, A., Théorie globale des connexions, 1958, Paris, Dunod.
Kobayashi, S., Nomizu, K., Introduction to differential Geometry, 1960, Acad. Press, N.Y.

(4) Trautman, A., 1970, Rep. Math. Phys., 1.
Kerner, R., 1968, Ann. Inst. H. Poincaré, 9, p. 143.
Kerbrat-Lunc, H., 1974, Ann. Inst. H. Poincaré, 21(4), p.333.
Cho, Y.M., 1975, Journ. Math. Phys., 1975, 16(10), p. 2029.
Rayski, J., 1965, Acta Phys. Polon., 27, p. 89.

(5) Corwin, L., Ne'eman, Y., Sternberg, S., 1975, Rev. Mod Phys., 647, p. 573.
Baha Balantekin, A., Bars, I., 1981, J. Math. Phys., 22 , p. 1149.

(6) Kerner, R., Group theoretical methods in Phys., 1977, ed. Sharp Kolman, Acad. Press, N.Y.
Kerner, R., da Silva Maia, E.M., 1983, Journ. Math. Phys., 24(2), p. 361.
Da Silva Maia, E.M., 1981, Lett. Math. Phys. 5 (6).

THE CONTINUITY OF COMPUTING CONNECTIONS FROM CURVATURES, AND OF DIVIDING SMOOTH FUNCTIONS

Mark Alan Mostow[1] and Steven Shnider[2]

(1) Mathematics Department, North Carolina State University, Raleigh, N.C. 27650 U.S.A.

(2) Mathematics Department, Ben-Gurion University of the Negev, B'er-Sheva`, Israel; on leave from Mathematics Department, McGill University, Montreal, P.Q., Canada

ABSTRACT: For a principal bundle with semi-simple structure group over a smooth four-dimensional base manifold, the set of connections (gauge potentials) A which are uniquely determined by their curvature (gauge field or field strength) F is generic in the set of all potentials, endowed with the Whitney C^∞ topology. The operator taking each such field F to its potential A is not continuous, however, nor are its restrictions to certain open dense subspaces. The relation of this problem to the question of whether the quotient of two smooth real-valued functions depends continuously on the numerator and denominator is examined.

In the functional integral approach to gauge field theory [GliJ], the space A of connections (gauge potentials) of a principal G-bundle P ⟶ M plays a key role. The quotient space A/G of gauge equivalence classes of gauge potentials [here G is the group of gauge transformations (automophisms of P ⟶ M over the identity map of M)] parametrizes the distinct physical states possible. By analogy with Feynman path integrals, one tries to define and compute various functional integrals over A/G. The work of Palais [Pal], Singer [Si-1,2], and others has shown that A/G

S. Sternberg (ed.), Differential Geometric Methods in Mathematical Physics, 45–54.

has a very rich geometric structure. It seems likely that as gauge field theory develops, more and more of this structure will be seen to have physical significance.

It has been pointed out by Halpern [Hal-1,2,3] that for some physical purposes it would be advantageous to work instead with the space B of curvatures (gauge fields or field strengths) and its quotient B/G. Mathematically, B/G has the advantage that G acts tensorially on B but not on A. To relate these two approaches to gauge field theory one must study the curvature operator $F: A \longrightarrow B$, which maps a connection A to its curvature

$$F = dA + (1/2)[A \wedge A],$$

and the induced operator

$$F': A/G \longrightarrow B/G.$$

An example of Wu and Yang [WuY] showed that F' is not a one-to-one map when G = SO(3). (F' is always onto by construction; it is a one-to-one correspondence when G is abelian and M is simply connected). Hence for non-abelian gauge field theories there are non-trivial problems, called <u>field copy problems</u>, of finding those curvatures (resp. elements of B/G) which are <u>copied</u>, i.e., are the image under F (resp. F') of more than one connection (resp. element of A/G). Various geometric, algebraic, and analytic criteria have been shown to give necessary and/or sufficient conditions on a connection for its curvature to be copied in one or both of these senses, that is, modulo gauge equivalence or not (see, for example, [Ati], [Cal], [Dao], [DesD], [DesT], [DesW], [GuY], [Hal-1,2,3], [Mos], [MosS-1,2], [Ro], [So], [Wei], [WuY]).

Despite the existence of field copies, the curvatures which are <u>not</u> copied are generic in a case of physical interest. More precisely:

Theorem 1. If dim M = 4 and G is a (finite-dimensional) semi-simple Lie group, then *A* contains generic (in fact, open and dense) subsets (in the Whitney C^{∞} topology, as defined in [GolG]) which are mapped in a one-to-one manner by *F* into *B*. Thus under these hypotheses, a generic connection A is determined uniquely by its curvature F.

Proof. See [MosS-1] and Theorem 1′ below.

Remark. This result has had the status of a "folk theorem" among physicists for several years, but to our knowledge, the first rigorous published proof of it is the one we gave recently in [MosS-1]. It was pointed out in several places (see, for example, [DesT], [DesW], [Hal-1]) that when the gauge group G is SU(2) or SO(3) and dim M = 4, the curvature determines the connection uniquely at each point of M, in a case called "generic". The precise topological description of the set of such connections and the extension from a pointwise to a global result on M were left unspecified. In contrast, Theorem 1 is valid for all semi-simple Lie groups G, and the term "generic" is used there in the rigorous global sense of [GolG] rather than the pointwise sense.

In light of Theorem 1, we may ask:

Question. Do there exist generic (or open dense) subspaces $A' \subset A$ which are (topologically) embedded by *F*, that is, mapped by *F* in a one-to-one and homeomorphic manner onto their image in *B*? In other words, can we choose a generic subspace *A′* of *A* so that $F|A'$ is one-to-one and $(F|A')^{-1}$ is continuous?

If there are, we shall be able to extend Theorem 1 to the statement that generically, a connection depends uniquely and continuously on its curvature (under the hypotheses of the theorem).

The largest possible candidate for such a subspace *A′* is

$$A^* = (\text{def.})\ \{A \in A \mid \text{no other connection has the same curvature}\}.$$

As we shall see soon, however, though A^* contains open

dense subsets of A and $F|A^*$ is one-to-one, the inverse map $(F|A^*)^{-1}$ is not continuous. We are forced, therefore, to consider subspaces of A^*.

Our approach will be inspired by a construction used by several authors (see, for example, [Cal], [DesT], [DesW], [Hal-1], and [Ro]), in which the Bianchi identity is used to find a sufficient condition for the non-existence of field copies (in the case dim M = 4). In the notation of differential forms, the Bianchi identity can be written

$$dF = [F \wedge A] = ad_F(A),$$

where ad_F denotes the map from **g**-valued 1-forms to 3-forms (**g** =(def.) Lie algebra of G) defined by wedging and bracketing with F. After choosing a local gauge for the bundle, local coordinates for M, and a basis for **g**, we can represent ad_F at each point $x \in M$ by an ordinary square matrix of dimension $(4 \dim G) \times (4 \dim G)$. If ad_F is invertible at a point $x \in M$, that is, if $\det(ad_F) \neq 0$ at x, then the connection A is determined uniquely by the curvature F at x by the formula

$$A = (ad_F)^{-1}(dF).$$

Globally, a sufficient condition for F to determine A uniquely is for the zero set of det ad_F to be nowhere dense in M, or equivalently, for the set

$$\{x \in M \mid \det ad_F(x) \neq 0\}$$

to be dense in M. Let A^{**} denote the set of all connections A satisfying this global condition. Thus $A^{**} \subset A^*$; in fact, an example in [MosS-2] shows that A^{**} is a proper subset of A^*.

Since the zero set of det ad_F is relevant to the curvature/connection problem, it is reasonable to guess that the behavior of det ad_F near its zero set may also be important. We shall therefore consider the subspaces

$$A_k = (\text{def.}) \{A \mid j^k(\det ad_F)(x) \neq 0 \text{ for all } x \in M\},$$
$$k=0,1,2,\ldots,\infty.$$

Here j^k denotes the k-jet, and the condition on A is equivalent to the non-vanishing, at each point $x \in M$, of some derivative of det ad_F of order $\leq k$.

To summarize, we have now defined a hierarchy of subspaces

$$A_0 \subset \ldots \subset A_k \subset \ldots \subset A_\infty \subset A^{**} \subset A^* \subset A .$$

We can now state a refined version of Theorem 1.

Theorem 1′. If dim M = 4, G is a semi-simple Lie group, and $k \geq 8$ dim G, then A_k is open and dense in A (in the Whitney C^∞ topology).

Remark. A much lower value of k probably works. On the other hand, A_0 is not dense, since it contains only connections for which det ad_F takes all positive or all negative values on each connected component of M. If G = SU(2) or SO(3), then $A_1 = A_0$, since det ad_F is then the square of a smooth function of F, by the formula of Wu and Yang quoted in [DesT]. (The point is that if $f = h^2$ and d_1 is a partial derivative operator, then $d_1 f = 2h d_1 h$, which vanishes on the zero set of f. Therefore, $j^1 f(x) \neq 0$ if and only if $j^0 f(x) \neq 0$.) Hence A_1 is not dense either in this case.

Proof of theorem. See [MosS-1]. The proof uses transversality theory, properties of algebraic sets, and the structure theory of semi-simple Lie algebras.

The spaces A_k with $k \geq 8$ dim G provide us with infinitely many candidates for A' in the Question above. The following result gives a partial negative answer to the Question.

Theorem 2. If dim M = 4, G = SU(2) or SO(3), and $k \geq 28$, then $(F|A_k)^{-1}$ is not continuous. That is, there exist connections A, $A_1 \in A_{28}$, with curvatures F, resp. F_1, such that $F_1 \to F$ in B but $A_1 \not\to A$ in A, in the respective Whitney C^∞ topologies.

Corollary. $(F|A^*)^{-1}$ and $(F|A^{**})^{-1}$ are not continuous.

Proof of theorem. See [MosS-1].

Discussion. The conditions on k in Theorems 1′ and 2 are probably far from the strongest possible. In any case, we have not ruled out the existence of an integer k for which A_k is open and dense and $(F|A_k)^{-1}$ is continuous. Nonetheless, we have come to believe that the spaces A_k are not the best candidates for A' in the

Question. In a different direction, we have outlined a proof that a certain space $A^\wedge$ does satisfy the conditions on A' in the Question. The space $A^\wedge$ contains the connections A whose curvatures F are transversal to all strata of a certain Whitney stratification of the zero set of det ad_* (regarded locally as a polynomial function on the Cartesian product $\mathbf{g}^6$, the space in which curvatures take their values).

RELATION TO DIVISION OF FUNCTIONS

In the process of studying the dependence of connections on curvatures, we were led to consider an interesting general problem concerning the continuity of the operation of dividing functions. The division problem arises as follows. Let C_F denote the transpose of the matrix of cofactors of ad_F. Then we can write

$$A = ad_F^{-1}(dF) = (C_F(dF)) \; / \; (\det ad_F).$$

Strictly speaking, this quotient makes sense only at points $x \in M$ where the denominator is not zero. But if $A \in A^{**}$, then the zero set of the denominator is nowhere dense, so that the value of the quotient is unambiguously determined at all points x, by continuity.

Suppose the quotient $h = f/g$ of two functions $f = gh$ and g (all functions here are C^∞ real- or complex-valued functions on a finite-dimensional manifold) were known to depend continuously on the pair (f,g), in the Whitney C^∞ topology. Then, since C_F and det ad_F depend continuously on F, the above formula would show that A depends continuously on F (for A in A^{**}). At first we naively conjectured that division of functions is continuous at all pairs (f=hg,g) for which g never vanishes to infinite order. Had this been true, it would have followed immediately that $(F|A_\infty)^{-1}$ is continuous. In fact, however, that conjecture was false, as can be shown by a counterexample involving the denominator $g(x,y) = x^2 + \exp(-1/y^2)$. [We found that example on our own, but later found that the same function g had been used by Malgrange [Mal] to answer an equivalent question about closed ideals in rings of smooth functions.] We exploited this counterexample to the continuity of division of functions to construct an example showing that $(F|A_{2\theta})^{-1}$ is discontinuous.

It appears that though much is known about the continuity of the quotient h as a function of the numerator f=hg for fixed denominator g (see, for example, [Ati], [Ehr-I,II,III,IV], [Hor], [Loj], [Mal], [Tou], [Whi]) the question of the joint continuity of the quotient as a function of the numerator and denominator has not been studied. (The continuity part of the Mather Division Theorem [Mat] sounds relevant to both of our continuity questions, but is, in fact, applicable to neither. To see this, observe that Mather's result (combined with the Malgrange Preparation Theorem [GolG]) holds for division by all functions which vanish to at most finite order, including the function $x^2 + \exp(-1/y^2)$ discussed above, division by which is discontinuous.) Moreover, a simple example will show that the conditions guaranteeing joint continuity of division must be stronger than those guaranteeing separate continuity. Let t be a real parameter, and let $f_t(x) = t$, $g_t(x) = t^2+x^2$. By classical results of Hörmander [Horm] and of Łojasiewicz [Loj], or by direct computation, division by g_t is continuous for each fixed t. But division of functions is not jointly continuous at the pair (f_o,g_o), since as $t \to 0$, $f_t \to f_o = 0$ and $g_t \to g_o = x^2$, but at $x = 0$, $h_t = f_t/g_t = t/t^2$ converges to ∞ and not to $h_o = f_o/g_o = 0$. Actually, the preceding statements are literally true only if we use the Fréchet C^∞ topology instead of the Whitney C^∞ topology, but the result is "locally" true in the Whitney C^∞ topology, too, and one can extend this "germ" of a counterexample to a global counterexample by suitable use of a cutoff function.

We have proved some results on the joint continuity of division. We plan to publish these soon, together with their consequences for curvature/connection problems, and in particular, a proof that the space $A^\wedge$ described above does indeed satisfy the conditions on A' in the Question.

REFERENCES

[Ati] Atiyah,M.F., *Resolution of singularities and division of distributions.* Commun. on Pure and Appl. Math. 23 (1970), pp.145-150.

[Cal] Calvo,M., *Connection between Yang-Mills potentials and their field strengths.* Phys. Rev. D 15 (1977), pp.1733-1735.

[Dao] Dao-xing,X., *On field strengths and gauge potentials of Yang-Mills' fields.* Scientia Sinica 20 (1977), pp.145-157.

[DesD] Deser, S., Drechsler, W., *Generalized gauge field copies.* Phys. Lett. 86B (1979), pp.189-192.

[DesT] Deser, S., Teitelboim, C., *Duality transforms of Abelian and non-Abelian gauge fields.* Phys. Rev. D 13 (1976), pp.1592-1597.

[DesW] Deser, S., Wilczek, F., *Non-uniqueness of gauge-field potentials.* Phys. Lett. 65B (1976), pp.391-393.

[Ehr-I] Ehrenpreis, L., *Solutions of some problems of division. I.* Am.J.Math. 76 (1954), pp.883-903.

[Ehr-II] ----------,----------. II. Am.J.Math. 77 (1955), pp.286-292.

[Ehr-III] ----------,----------. III. Am.J.Math. 78 (1956), pp.685-715.

[Ehr-IV] ----------,----------. IV. Am.J.Math. 82 (1960), pp.522-588.

[GliJ] Glimm,J., Jaffe, A., *Quantum Physics: A Functional Integral Point of View.* Springer, New York, 1981.

[GolG] Golubitsky, M., Guillemin, V., *Stable Mappings and their Singularities.* New York: Springer-Verlag, 1973.

[GuY] Gu, C.-H., Yang, C.-N., *Some problems on the gauge field theories, II.* Sci. Sin. 20 (1977), pp.47-55.

[Hal-1] Halpern, M. B., *Field strength formulation of quantum chromodynamics.* Phys. Rev. D 16 (1977), pp.1798-1801.

[Hal-2] Halpern, M. B., *Field strength copies and action copies in quantum chromodynamics.* Nucl. Phys. B 139 (1978), pp.477-489.

[Hal-3] Halpern, M. B., *Field strength and dual variable formulations of gauge theory.* Phys. Rev. D 19 (1979), pp.517-530.

[Horm] Hörmander, L., *On the division of distributions by polynomials.* Arkiv för Matematik 3 (1958), pp.555-568.

[KobN] Kobayashi, S., Nomizu, K., *Foundations of Differential Geometry, Part 1.* New York: Interscience, 1963.

[Loj] Łojasiewicz, S., *Sur le problème de la division.* Studia Math. 18 (1959), pp.87-136.

[Mal] Malgrange, B., *Ideals of Differentiable Functions.* Oxford Univ. Press, London, 1966.

[Mat] Mather, J.N., Stability of C^∞ *mappings: II. Infinitesimal stability implies stability.* Annals of Math. 89 (1969), pp.254-291.

[Mos] Mostow, M. A., *The field copy problem: to what extent do curvature (gauge field) and its covariant derivatives determine connection (gauge potential)?* Comm. Math. Phys. 78 (1980), pp.137-150.

[MosS-1] Mostow, M. A., Shnider, S., *Does a generic connection depend continuously on its curvature?* To appear in Communications in Mathematical Physics.

[MosS-2] Mostow, M. A., Shnider, S., *Counterexamples to some results on the existence of field copies.* To appear in Communications in Mathematical Physics.

[Pal] Palais, R., Letter to I. Singer (concerning the differential geometry of *A/G*), (1977).

[Ro] Roskies, R., *Uniqueness of Yang-Mills potentials.* Phys. Rev. D 15 (1977), pp.1731-1732.

[Si-1] Singer, I. M., *Some remarks on the Gribov ambiguity.* Comm. Math. Phys. 60 (1978), pp.7-12.

[Si-2] Singer, I.M., *The geometry of the orbit space for nonabelian gauge theories.* Physica Scripta 24 (1981), pp.817-820.

[So] Solomon, S., *On the field strength-potential connection in non-abelian gauge theory.* Nucl. Phys. B 147 (1979), pp.174-188.

[Tou] Tougeron, J.-C., *Idéaux de fonctions différentiables.* Springer, New York, 1972.

[Wei] Weiss, N., *Determination of Yang-Mills potentials from the field strengths.* Phys. Rev. D 20 (1979), pp.2606-2609.

[Whi] Whitney, H., *On ideals of differentiable functions.* Amer. J. Math., 70 (1948), pp.635-658.

[WuY] Wu, T. T., Yang, C. N., *Some remarks about unquantized non-abelian gauge fields.* Phys. Rev. D 12 (1975), pp.3843-3844.

GAUGE THEORIES ON HOMOGENEOUS MANIFOLDS

A. Pérez-Rendón and D.H. Ruipérez

Department of Mathematics, University of Salamanca, Spain

A definitive (if possible) gauge formulation of gravitation necessitates the previous resolution of a double problem:

a) What is the gauge group of gravitation?

b) What is the Lagrangian of the gravitational field?

Since none of the solutions proposed up to the present seems to be wholly convincing, we feel it of interest to generalize the Cartan-Kibble-Sciama theory and that of Hehl et al. for a Lie group G — fibered on a weakly reductive homogeneous manifold $X = G/H$ — and to determine, in each case, the conditions that the Lagrangians of its corresponding gauge fields should fulfil.

The results obtained are:

1) The C-K-S theory is *globally and intrinsically* formulable for *non trivial* principal bundles $p:G \longrightarrow X$. We also have:

Theorem 1: The Lagrangian of the material (or initial) field will be invariant for all the infinitesimal automorphisms of $p:G \longrightarrow X$ if and only if it is invariant by the Lie algebra of the group G.

2) Hehl's theory is formulable for trivial principal bundles $p:G \longrightarrow X$, X being an open subset of some $\mathbf{R}^n$.

INTRODUCTION

Many attempts have been made since 1956 to construct a gauge theory of gravitation and several generalizations of the Yang-Mills-Utiyama theory of interaction between classical fields have ensued from them.

Among such generalizations, the most important from a mathematical point of view are: the Cartan-Kibble-Sciama Theory

S. Sternberg (ed.), Differential Geometric Methods in Mathematical Physics, 55–74.

(C-K-S) [1,2] and the Poincaré gauge Field Theory (PGF) by Hehl et al. [3-5] .

Even though both theories differ in their mathematical formulation and in their consequences of physical order, the starting point is the same:

We assume that a classical field (matter field or initial field) is given on a vector bundle $\pi_1 : E \longrightarrow \mathbf{R}^4$ associated with the trivial principal bundle $p : P \longrightarrow \mathbf{R}^4$, whose structure group is the Lorentz group, P being the Poincaré group.

In this paper we have a double aim:

First, *to state both theories with the greatest possible generality.*

That is, we now suppose that E is a vector bundle associated with a principal bundle $p:G \longrightarrow X$ where G is a *Lie group,* $H \hookrightarrow G$ *a closed Lie subgroup* and the quotien manifold $X = G/X$ is a *weakly reductive homogeneous manifold* in the following sense:

Let $\mathcal{G}$ and $\mathcal{H}$ be the Lie algebras of G and H. There exists a direct summand $\mathcal{F}$ of $\mathcal{H}$ in $\mathcal{G}$ which is invariant under the adjoint representation of the elements of $\mathcal{H}$, i.e., $\mathrm{ad}(h)\mathcal{F} \subseteq \mathcal{F}$, for every $h \in \mathcal{H}$.

According to the work by Ne'eman and Regge [6,7] on which this paper is partially based, and the growing interest on supergroups and superspaces, we think it unnecessary to justify such a degree of generality when dealing with C-K-S and PGF theories.

Nevertheless, we shall point out one reason for doing so:

If we concentrate on the purpose that both theories were constructed for, i.e. giving a gauge formulation of gravitation, a problem which as yet remains unsolved is:

What is the gauge group of gravitation?

(See, for instance, Trautman [8]).

Thus, it is important to have both theories formulated in the most general case, instead of the very specific case of the Poincaré group.

We shall therefore assume hereafter that $p:G \longrightarrow X$ is a principal bundle, in general a non-trivial bundle, with a structure group $H \hookrightarrow G$ and whose base manifold $X = G/H$ is a weakly reductive homogeneous manifold.

These hypotheses yield two well-known results (cf. [9], Chapter II, Section 11) which will be used throughout this paper:

<u>Proposition 1</u>: Let $G \times \mathcal{G}/H$, $\mathrm{Ad}\, G$ and $\mathrm{Ad}\, \mathcal{F}$ denote the vector bundles associated to $p:G \longrightarrow X$ with standard fibers $\mathcal{G}$, $\mathcal{H}$ and $\mathcal{F}$.

There exists an isomorphism of vector bundles over X between $G \times \mathcal{G}/H$ and the bundle $T_H G$ of H-invariant fields on G, such that the following diagram

$$\begin{array}{ccccccccc} 0 & \longrightarrow & \mathrm{Ad}\, G & \longrightarrow & G \times \mathfrak{g}/H & \longrightarrow & \mathrm{Ad}\, F & \longrightarrow & 0 \\ & & \downarrow \wr & & \downarrow \wr & & & & \\ 0 & \longrightarrow & V_H G & \longrightarrow & T_H G & \longrightarrow & TX & \longrightarrow & 0 \end{array}$$

is commutative; $\mathrm{Ad}\, G \xrightarrow{\sim} V_H G$ denotes the natural isomorphism between the adjoint bundle and the bundle of vertical, H-invariant, vector fields on G.

Consequently, there exists a connection $\sigma_0 : TX \xrightarrow{\sim} \mathrm{Ad}\, F \hookrightarrow T_H G$, whose horizontal distribution is $\mathrm{Ad}\, F$.

Proposition 2: $T_H G$ is a trivial bundle with standard fiber $\mathfrak{g}$. The isomorphism of $C^\infty(X)$-modules

$$C^\infty(X) \otimes_{R} \mathfrak{g} \xrightarrow{\sim} \Gamma(T_H G)$$

is given by right G-invariant vector fields, corresponding to a basis of $\mathfrak{g}$.

Our second aim, in this paper, is *to give a determination of all possible Lagrangians* V *of the gauge field associated with the matter field* E.

The reasons for studying — also with great generality — this problem are the following:

One of the most controversial points of the different attempts to derive a theory of gravitation from a gauge principle is the choice of a Lagrangian for the gravitational field.

Furthermore, according to whether we apply the C-K-S theory or the PGF theory, the possible Lagrangians for the gravitational field will fulfil different systems of differential equations.

We therefore think it important to fix the equations which the Lagrangians of a generalized gauge field should fulfil.

The paper is organised as follows. In Section 1, we shall discuss the C-K-S theory with special emphasis on the following points:

1) Definition of a principle of minimal interaction.
2) Characterization of the Lagrangian of the matter field.
3) Determination of the Lagrangian of the associated gauge field.
4) Demonstration of the gauge invariance of the principle of minimal interaction.

Section 2 deals with the PGF theory following the same scheme.

1. CARTAN-KIBBLE-SCIAMA THEORY

1.1. The structure form $\bar{\vartheta}$

1.1.1. Let $\pi_1 : E \longrightarrow X$ be a vector bundle associated with G with standard fiber F. We define on E the bundle

$$\bar{\pi}_1 : \bar{E} = \mathrm{Hom}(\pi_1^* TX, VE) \longrightarrow E$$

where $\pi_1^* TX$ is the pull-back of TX on E and VE is the bundle of vertical tangent vectors to E.

$\bar{E}$ is the vector bundle associated with the affine bundle

$$J^1E \longrightarrow E \ .$$

In order to define a classical field on E one must give on J^1E (or on $\bar{E}$) a Lagrangian density form Λ and a structure 1-form ϑ, with values in the module $\Gamma(\bar{\pi}_1^* VE)$.

On J^1E there is a canonical structure form. Let us define it:

Let x be an arbitrary point of X, U an open neighbourhood containing x and $s : U \longrightarrow E$ a local section of E over U.

If $j_x^1 s$ is the one jet of s at x and $\bar{D}$ is a tangent vector to J^1E at $j_x^1 s$, then

$$\vartheta_{j_x^1 s}(\bar{D}) = d^V s_x(D)$$

where $D = \bar{\pi}_{1*} D$ and $d^V s_x$ is the vertical differential of s at x, i.e., the projector of $TE_{s(x)}$ on $VE_{s(x)}$ given by the exact sequence

$$0 \longrightarrow TX_x \xrightarrow{\ ds_x\ } TE_{s(x)} \xrightarrow{\ d^V s_x\ } VE_{s(x)} \longrightarrow 0 \ .$$

Obviously, if $s, s' \varepsilon j_x^1 s$, their vertical differential at x, $d^V s_x$ and $d^V s'_x$, agree.

However, the structure form on $\bar{E}$ is not unique.

Given a connection σ on E, we can define on $\bar{E}$ a structure form ϑ^σ in the following way:

$$\vartheta^\sigma_{\bar{y}}(\bar{D}) = D_y - \sigma_x(D') + \psi(D'_y)$$

$\bar{D}$ is any tangent vector to $\bar{E}$ at $\bar{y} = (y, \psi)$; $\psi \varepsilon \mathrm{Hom}(\pi_1^* TX, VE)_y$, $x = \pi_1(y)$ and D'_x, D_y are the projections of $\bar{D}_{\bar{y}}$ on X and E, respectively.

1.1.2. Since E is a vector bundle associated to $p : G \longrightarrow X = G/H$, there exists a representation of $C^\infty(X)$-Lie algebras:

$$\mathrm{aut}(G) \simeq \Gamma(T_H G) \xrightarrow{\xi_1} \Gamma(T_{\pi_1} E)$$

with values on π_1-projectable vector fields and consistent with the projections on X :

$$\begin{array}{ccc} \mathrm{aut}(G) & \xrightarrow{\xi_1} & \Gamma(T_{\pi_1} E) \\ & {\scriptstyle p_*}\searrow \quad \swarrow{\scriptstyle \pi_{1*}} & \\ & \Gamma(TX) & \end{array}$$

so:

$$\mathrm{gau}(G) \simeq \Gamma(\mathrm{Ad}\ G) \xrightarrow{\xi_1} \Gamma(VE) \ .$$

If we compose the natural homomorphism of Lie algebras $\mathcal{g} \longrightarrow \mathrm{aut}(G)$, given by the right G-invariant vector fields, with ξ_1 , we obtain a representation of Lie algebras $\mathcal{g} \longrightarrow \Gamma(T_{\pi_1} E)$.

<u>Remark 1</u>: We write $\mathrm{aut}(G) \simeq \Gamma(T_H G)$ and $\mathrm{gau}(G) \simeq \Gamma(\mathrm{Ad}\ G)$ because $\Gamma(T_H G)$ and $\Gamma(\mathrm{Ad}\ G)$ are the "Lie algebras" of the groups:

$$\mathrm{Aut}(G) = \{\text{automorphisms of the principal bundle } G\}$$

$$\mathrm{Gau}(G) = \{\text{automorphisms of } p:G \longrightarrow X \text{ with a trivial action on } X\}$$

(see Sternberg [10]).

Let $U \subseteq X$ be a coordinate neighborhood, with a local coordinate system $(x_1,\ldots,x_n)$, such that $p^{-1}(U) \simeq U \times H$ and $\pi_1^{-1}(U) \simeq U \times F$.

Let $\{A_N\}$, $\{B_j\}$ and $(z_1,\ldots,z_m)$ be arbitrary bases of H , F and F^* respectively.

Each element A_N , B_j of the chosen basis of $\mathcal{g}$ gives arise to a section $\tilde{A}_N$, $\tilde{B}_j$ of the vector bundle $T_H(G|_U)$ — the corresponding H-invariant vector field — in such way that the $\{\tilde{A}_N\}$ span the vertical subspace and the $\{\tilde{B}_j\}$ span the horizontal subspace of the connection σ_0 .

On the other hand, we can associate to $\{A_N,B_j\}$ the basis $\{RA_N,RB_j\}$ of $\Gamma(T_H G)$, whose elements are the corresponding right G-invariant vector fields on G . Locally, we have,

$$RA_N = f_N^M \tilde{A}_M + f_N^j \tilde{B}_j$$

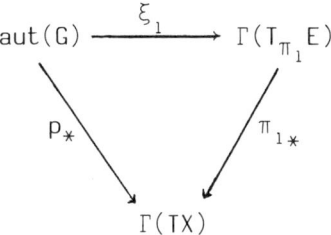

$$RB_j = f_j^M \tilde{A}_M + f_j^{\ell} \tilde{B}_{\ell}$$

where the $f \varepsilon C^{\infty}(U)$.

If $p_* \tilde{B}_j : = g_{ij} \frac{\partial}{\partial x_i}$, we can write

$$\tilde{B}_j : = g_{ij} \frac{\partial}{\partial x_i} + g_{Mj} \tilde{A}_M$$

the g being also functions of $C^{\infty}(U)$.

So, any $\gamma \varepsilon aut(G)$ can be, locally, given by

$$\gamma = f_i \frac{\partial}{\partial x_i} + g_M \tilde{A}_M$$

and the computation of $\xi_1(\gamma)$ reduces to $\xi_1(\tilde{A}_M)$.

Let q be the projection of $G \times F$ on the quotient space $E = G \times F/H$, e a vector of F and q_e the mapping of G into E given by

$$q_e(u) = q(u,e) : = \overline{(u,e)}$$

for $u \varepsilon G$.

We define ξ_1 :

$$\xi_1(\tilde{A}_M)_{\overline{(u,e)}} = q_{e*}((\tilde{A}_M)_u)$$

and we obtain

$$\xi_1(\tilde{A}_M) = a_{hk}^M z_h \frac{\partial}{\partial z_k}$$

where $a_{hk}^M \varepsilon \mathbf{R}$.

<u>1.1.3.</u> Let V be an open coordinate neighborhood in $\bar{E}$ with local coordinates (x_i, z_j, p_{ij}) such that

$$p_{ij}(y) = -[\vartheta_{\bar{y}}^{\sigma_0}(\frac{\partial}{\partial x_i})]z_j$$

for every $\bar{y} \varepsilon V$.

The definition of ϑ^{σ_0} yields, locally:

$$\vartheta^{\sigma_0} = (dz_j - p_{ij}dx_i) \circ \frac{\partial}{\partial z_j}$$

If σ is an arbitrary connection on G , the form ϑ^{σ} can be written in the former coordinate system as:

$$\vartheta^{\sigma} = (dz_j - (p_{ij} + \Gamma_{iN}(x) a_{kj}^N z_k) dx_i) \circ \frac{\partial}{\partial z_j}$$

where $\Gamma_{iN}(x)a^N_{kj}z_k dx_i \circ \frac{\partial}{\partial z_j} = \Gamma_{iN}(x)dx_i \circ \xi_1(\tilde{A}_N)$ is the local expression of the (VE)-valued 1-form on X, associated with the connections (σ, σ_0).

Let $\pi_2 : L(X) \longrightarrow X$ be the *bundle of linear frames over* X, whose fiber over $x \varepsilon X$ is the set of linear isomorphisms

$$\varphi_x : \mathbf{R}^n \longrightarrow T_x(X) .$$

We assume that $U = \pi_1 \circ \bar{\pi}_1(V)$ and that $\varphi : U \longrightarrow L(X)$ is a local section of $L(X)$ given by

$$\varphi : x \longrightarrow \varphi_x .$$

Each local section φ *of* $L(X)$ *allows us to define a local coordinate system* $(x_i, z_j, \bar{p}_{ij})$ on V in the following way:

Take an arbitrary basis $(e_1, \ldots, e_n)$ of $\mathbf{R}^n$ and let $(D_k)_x = \varphi_x(e_k) = e^j{}_k(x) \frac{\partial}{\partial x_i}$. If $\bar{y}$ is a point of V such that $x = \pi_1 \circ \bar{\pi}_1(\bar{y})$,

$$\bar{p}_{ij}(\bar{y}) = -[\vartheta^{\sigma_0}_{\bar{y}}(D_i)]z_j = e^k{}_i(x)p_{kj}(\bar{y}) .$$

Consequently, if $(e_k{}^{\ell}(x))$ denotes the inverse matrix of $(e^k{}_{\ell}(x))$, then

$$e_i{}^{\ell}(x)e^j{}_{\ell}(x) = e^i{}_{\ell}(x)e_j{}^{\ell}(x) = \delta_{ij}$$

and we also have

$$p_{kj}(\bar{y}) = e_k{}^{\ell}(x)\bar{p}_{\ell j}(y) .$$

Using the new coordinate system,

$$\vartheta^{\sigma} = (dz_j - (e_i{}^{\ell}(x)\bar{p}_{\ell j} + \Gamma_{iN}(x)a^N_{kj}z_k)dx_i) \circ \frac{\partial}{\partial z_j} \qquad \text{(A)}$$

where $e_i{}^{\ell}(x)$, $\Gamma_{iN}(x) \varepsilon C^{\infty}(U)$.

1.1.4. Let K be the *vector bundle of connections on* $p : G \longrightarrow X$, i.e.,

$$K = \mathrm{Hom}(TX, \mathrm{Ad}\, G) \xrightarrow{\Pi_3} X ,$$

and let Y, $\bar{Y}$ denote the fibred products:

$$Y = E \times_X L(X) \times_X K \xrightarrow{\pi} X \; ; \; \bar{Y} = \bar{E} \times_X J^1 L(X) \times_X J^1 K \xrightarrow{\bar{\pi}} X .$$

We may define a structure form $\bar{\vartheta}$ on $\bar{Y}$ as follows:

For every $z = (y, j^1_x\varphi, j^1_x(\sigma-\sigma_0)) \varepsilon \bar{Y}$ and for every tangent vector to $\bar{Y}$ at $\bar{z}$, $\bar{D}_{\bar{z}} = ((\bar{D}_1)_{\bar{y}}, (\bar{D}_2)_{j^1_x\varphi}, (\bar{D}_3)_{j^1_x(\sigma-\sigma_0)})$, we have

$$\bar{\vartheta}_{\bar{z}}(\bar{D}) = \vartheta^{\sigma}_{\bar{y}}(\bar{D}_1) + \vartheta'_{j^1_x\varphi}(\bar{D}_2) + \vartheta''_{j^1_x(\sigma-\sigma_0)}(\bar{D}_3)$$

where ϑ', ϑ'' are the structure forms canonically defined on $J^1L(X)$ and J^1K.

Remark 2: There exists a vector bundle homomorphism of $T_{\pi_1}E \times_{T(X)} T_{\pi_2}L(X) \times_{T(X)} T_{\pi_3}K$ into $T_{\pi}Y$ whose local description is

$$(D_1, D_2, D_3) \longrightarrow D + D'_1 + D'_2 + D'_3$$

with $\pi_{i*}(D_i) = D$, for $i = 1,2,3$, $D'_i = D_i - D$ and $T_{\pi_i}E_i$ being the bundle on X of π_i-projectables vector fields on E_i.

1.2. Local expression of $\bar{\vartheta}$

Let V be an open neighborhood of Y whose projection on X is an open coordinate neighborhood U, such that $E|_U$, $L(X)|_U$ and $K|_U$ are trivial bundles.

Let $(x_1, \ldots, x_n)$ be a local coordinate system on U. We can define another local coordinate system $(x_i, z_j, e_\ell{}^h, \Gamma_{kN}, p_{ij}, H^h_{i\ell}, B_{ikN})$ on V in the following way:

$$e_\ell{}^h(x, \varphi_x) = e^*_h[\varphi_x^{-1}(\frac{\partial}{\partial x_\ell})] \quad ; \quad \forall (x, \varphi_x) \varepsilon \pi_2^{-1}(U)$$

$$\Gamma_{kN}(x, (\sigma-\sigma_0)_x) = A^*_N((\sigma-\sigma_0)_x \frac{\partial}{\partial x_k}) \quad ; \quad \forall (x, (\sigma-\sigma_0)_x) \varepsilon \pi_3^{-1}(U) \, .$$

Finally, for every $\bar{z} = (\bar{y}, j^1_x\varphi, j^1_x(\sigma-\sigma_0))$ we have

$$p_{ij}(\bar{z}) = -[\vartheta^{\sigma_0}_y(\varphi_x(\frac{\partial}{\partial x_i}))]z_j \quad ; \quad H_i{}^h{}_\ell(\bar{z}) = -[\vartheta'_{j^1_x\varphi}(\frac{\partial}{\partial x_i})]e_\ell{}^h \, ;$$

$$B_{iKN}(\bar{z}) = -[\vartheta''_{j^1_x(\sigma-\sigma_0)}(\frac{\partial}{\partial x_i})]\Gamma_{kN} \, .$$

In such coordinate system we can write:

$$\bar{\vartheta} = [dz_j - (e_i{}^\ell p_{\ell j} + \Gamma_{iN} a^N_{kj} z_k)dx_i] \circ \frac{\partial}{\partial z_j} +$$

$$+ (de_\ell{}^h - H_i{}^h{}_\ell dx_i) \circ \frac{\partial}{\partial e_\ell{}^h} + (d\Gamma_{kN} - B_{ikN}dx_i) \circ \frac{\partial}{\partial \Gamma_{kN}} \, . \text{(B)}$$

Between the form $\bar{\vartheta}$ defined on $\bar{Y}$ and the set $\{\vartheta^\sigma\}$ of forms defined on $\bar{E}$ there the following relation:

Proposition 1 (cf. [11]): Every local section (σ,φ) of $L(X)\times_X K$ induces a local section $i_{\sigma,\varphi}$ of the bundle $\bar{Y}\longrightarrow\bar{E}$, such that

$$\bar{\vartheta}|_{\mathrm{Im}\, i_{\sigma,\varphi}} = \vartheta^\sigma \quad \text{"written in the reference } \varphi\text{"}$$

(see formula A).

Next, we shall define on $\bar{Y}$ endowed with the structure form $\bar{\vartheta}$, a *minimal coupling* between a *matter field* and a *gauge field* given on $E\times_X L(X)$ and $L(X)\times_X K$, respectively.

1.3. Lagrangian density of the matter field

1.3.1. The soldering form on $L(X)$

Let η be the 1-form on $L(X)$ with values in the trivial bundle $L(X)\times\mathbf{R}^n$, given by the formula

$$\eta_{\varphi_x}(D) = \varphi_x^{-1}(\pi_{2*}D_{\varphi_x})$$

In the local coordinate system $(x_i, e_h{}^k)$, η can be written as:

$$\eta = \eta_j \cdot e_j = (e_i{}^j dx_i)\cdot e_j .$$

Moreover, on $L(X)$ we can define a n-form $(n = \dim X)$

$$\omega = \eta_1 \wedge \ldots \wedge \eta_n$$

which, in coordinates, is

$$\omega_{\varphi_x} = \mathrm{Det}(\varphi_x^{-1})dx_1 \wedge \ldots \wedge dx_n .$$

Every differentiable function $L:\bar{E}\longrightarrow\mathbf{R}$ gives rise to a Lagrangian density $\Lambda = L\omega$ on $\bar{Y}$.

We wish Λ to be the Lagrangian density of a matter field, so we must find suitable conditions for L.

1.3.2. We first *construct a representation* ζ of $\mathrm{aut}(G)$ in $\Gamma(T_\pi Y)$ using the three following representations:

1. $\xi_1:\mathrm{aut}(G)\longrightarrow\Gamma(T_{\pi_1}F)$; see 1.1.2.
2. $\xi_2:\mathrm{aut}(G)\longrightarrow\Gamma(T_{\pi_2}L(X))$, defined as:

$$\gamma\longrightarrow\xi_2(\gamma) = \bar{D}$$

where $D = p_*\gamma$ and $\bar{D}$ is the unique vector field on $L(X)$ being π_2-projectable on D and such that $L_{\bar{D}}\,\eta = 0$.

Locally, if $D = f_i(x)\dfrac{\partial}{\partial x_i}$, $\bar{D} = f_i\dfrac{\partial}{\partial x_i} - \dfrac{\partial f_k}{\partial x_j}e_k{}^{\ell}\dfrac{\partial}{\partial e_j{}^{\ell}}$.

3. $\xi_3 : \mathrm{Aut}(G) \longrightarrow \Gamma(T_{\pi_3}K)$.

In order to define this third representation one must proceed as follows (P.L. García [12]).

Since H acts freely on J^1G on the right, we can write:

$$\begin{array}{ccc} J^1G & \xrightarrow{\ \bar{p}\ } & G \\ \Big\downarrow & & \Big\downarrow p \\ J^1G/H & \longrightarrow & X \end{array}$$

Now, $J^1G/H \longrightarrow X$ is isomorphic with the bundle $\pi_3 : K \longrightarrow X$, and $J^1G \simeq \pi_3^*G$ so we also have:

$$\begin{array}{ccc} \pi_3^*G & \xrightarrow{\ \bar{p}\ } & G \\ p'\Big\downarrow & & \Big\downarrow p \\ K & \xrightarrow{\ \pi_3\ } & X \end{array}$$

where $p' : \pi_3^*G \longrightarrow K$ is a principal H-bundle.

Proposition 2: The structure form on J^1G is a *connection 1-form* (we shall call such connection the *canonical connection* on π_3^*G).

In fact, the structure form on J^1G is $\Gamma(\bar{p}^*VG)$-valued, but, on account of the canonical isomorphism between VG and $G\times\mathcal{H}$, we can also regard it as a $\mathcal{H}$-valued 1-form.

Construction of ξ_3

If $\gamma \in \mathrm{aut}(G)$ and $\bar{\gamma}$ is the 1-jet prolongation of γ to J^1G, then:

$$\xi_3(\gamma) := D_\gamma = p'_*(\bar{\gamma})$$

Moreover, if Ω is the 2-form on K, with values in $\Gamma(\pi_3^*\mathrm{Ad}\ G)$, which correspond to the curvature form of the canonical connection on π_3^*G, we have:

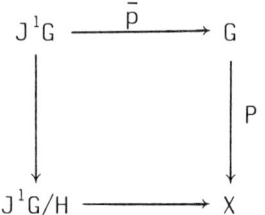

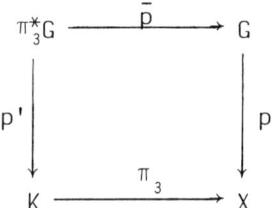

Proposition 3: There exists an element $\tilde{\gamma} \varepsilon \Gamma(\pi_3^* \mathrm{Ad}\ G)$ such that

$$i D_\gamma \cdot \Omega = d\tilde{\gamma}$$

and $d\tilde{\gamma}$ is calculated by means of the canonical connection.

Locally, if $\gamma = f_i(x) \frac{\partial}{\partial x_i} + g_M(x) \tilde{A}_M$ then $D_\gamma = f_i \frac{\partial}{\partial x_i} + (\frac{\partial g_N}{\partial x_i} - \Gamma_{kN} \frac{\partial f_k}{\partial x_i} + c^N_{AM} \Gamma_{iA} g_M) \frac{\partial}{\partial \Gamma_{iN}}$ and we can write $\tilde{\gamma} = (g_M - \Gamma_{kM} f_K) \tilde{A}_M$. The c^N_{AM} are the structure constants of H.

Remark 3: Let $\bar{\Gamma}$ be the subalgebra of $\Gamma(\pi_3^* \mathrm{Ad}\ G)$ whose elements are the sections s such that there exists a vector field D_s on K obeying the condition $i D_s \cdot \Omega = ds$. A Poisson bracket can be defined on $\bar{\Gamma}$:

$$\{s,s'\} = [s,s'] - \Omega(D_s, D_{s'}) \quad .$$

In particular, if $\gamma, \gamma' \varepsilon \mathrm{Ad}\ G$ and $D_\gamma = \xi_3(\gamma)$; $D_{\gamma'} = \xi_3(\gamma')$ we have

$$d\{\tilde{\gamma}, \tilde{\gamma}'\} = i[D_\gamma, D_{\gamma'}] \cdot \Omega = i D_{[\gamma,\gamma']} \cdot \Omega$$

Construction of the representation ζ *of* aut(G) *in* $\Gamma(T_\pi Y)$:

$$\begin{array}{ccc} & \Gamma(T \times E) \times_{\Gamma(TX)} \Gamma(T_{\pi_2} L(X)) \times_{\Gamma(TX)} \Gamma(T_{\pi_3} K) & \\ {\scriptstyle (\xi_1,\xi_2,\xi_3)} \nearrow & & \searrow \\ \mathrm{aut}(G) & \xrightarrow{\quad\quad\zeta\quad\quad} & \Gamma(T_\pi Y) \end{array}$$

making use of Remark 2.

Local expression

If $\gamma = f_i \frac{\partial}{\partial x_i} + g_M \tilde{A}_M$, then

$$\zeta(\gamma) = f_i \frac{\partial}{\partial x_i} + g_M a^M_{hk} z_h \frac{\partial}{\partial z_k} + (\frac{\partial g_N}{\partial x_i} - \Gamma_{kN} \frac{\partial f_K}{\partial x_i} - c^N_{MA} \Gamma_{iA} g_M) \frac{\partial}{\partial \Gamma_{iN}} - \frac{\partial f_k}{\partial x_i} e_k^{\ell} \frac{\partial}{\partial e_i^{\ell}}$$

1.3.3. Definition 1: We call a matter field on $\bar{Y}$, with Lagrangian density $\Lambda = L\omega$, G-*invariant* if $L_{\overline{\zeta(A)}}(L\omega) = 0$, for any $A \in g$, where $\overline{\zeta(A)}$ denote the $\bar{\vartheta}$-lifting of $\zeta(A)$ to $\bar{Y}$, i.e., $\overline{\zeta(A)}$ is the unique vector field on $\bar{Y}$, $\bar{\vartheta}$-projectable on $\zeta(A)$ and such that

$$L_{\overline{\zeta(A)}}\bar{\vartheta} = \varphi \circ \bar{\vartheta} .$$

(If we write $M = \Gamma(\bar{\pi}^* VY)$, then φ is a Hom(M,M)-valued function, the product $(\circ)$ is induced by the bilinear product $\mathrm{Hom}(M,M) \times M \longrightarrow M$ and the Lie derivative $L_{\zeta(A)}$ is calculated by using any derivation law ∇ in M).

Definition 2: A matter field on $\bar{Y}$, with Lagrangian density $\Lambda = L\omega$, is said to be invariant under all infinitesimal automorphisms of $p: G \longrightarrow X$, i.e., aut(G)-*invariant*, if $L_{\bar{D}}(L\omega) = 0$, where $\bar{D}$ denote the $\bar{\vartheta}$-lifting of D to $\bar{Y}$ for any $D \in \zeta(\mathrm{aut}(G))$.

Theorem 1: *A matter field* $(Y, L\omega)$ *is* aut(G)-*invariant if and only if it is* G-*invariant*.

Proof: Let $\gamma = \mu_M(x)(RA_M) + \lambda_j(x)(RB_j)$ be an element of aut(G). From the definitions of ω and ξ_2 it follows:

$$L_{\overline{\zeta(\gamma)}}(L\omega) = \overline{\zeta(\gamma)}(L)\omega + L\, L_{\overline{\zeta(\gamma)}}\omega =$$

$$= \overline{\zeta(\gamma)}(L)\omega + L\, L_{\overline{\zeta(\gamma)}}(\eta_1 \wedge \ldots \wedge \eta_n) = \overline{\zeta(\gamma)}(L)\omega.$$

A local computation gives:

$$\overline{\zeta(\gamma)}(L) = a_j \frac{\partial L}{\partial x_j} + f_M a^M_{hk} z_h \frac{\partial L}{\partial z_k} + \overline{\zeta(\gamma)}(p_{ij}) \frac{\partial L}{\partial p_{ij}}$$

if $\gamma = a_j \frac{\partial}{\partial x_j} + f_M \tilde{A}_M$.

But, using the definition of the $\bar{\vartheta}$-lifting of $\zeta(\gamma)$ to $\bar{Y}$, we obtain:

$$L_{\overline{\zeta(\gamma)}}\bar{\vartheta}_j = f_M a^M_{hj} \bar{\vartheta}_h$$

where $\bar{\vartheta}_j = dz_j - (e^{\ell}_i p_{\ell j} + \Gamma_{iN} a^N_{kj} z_k) dx_i$. So, $\overline{\zeta(\gamma)}(p_{ij}) =$ $= f_M a^M_{hj} p_{ih}$.

And thus, $L_{\overline{\zeta(\gamma)}}(L\omega) = \left[a_j \frac{\partial L}{\partial x_j} + f_M a^M_{hk}\left(z_h \frac{\partial L}{\partial z_k} + p_{ih}\frac{\partial L}{\partial p_{ik}}\right)\right]\omega =$

$$= \mu_M L_{\overline{\zeta(RA_M)}}(L\omega) + \lambda_j L_{\overline{\zeta(RB_j)}}(L\omega)$$

since the coefficients a_j, f_M depend linearly on γ.

1.4. Characterization of all possible Lagrangians V of the generalized gauge field $L(X)\times_X K$.

The homomorphisms ξ_2 and ξ_3 also induce a Lie algebra representation of $\mathrm{aut}(G)$ in the vector fields on $L(X)\times_X K$, projectable on X; we shall denote such representation by ζ'.

Definition 3: The Lagrangians of the generalized gauge field $L(X)\times_X K$ are functions $V: J^1L(X)\times_X J^1K \longrightarrow \mathbf{R}$ obeying the following law:

$$L_{\overline{\zeta'(\gamma)}}(V\omega) = 0$$

where, for any $\gamma\in\mathrm{aut}(G)$, $\overline{\zeta'(\gamma)}$ is the 1-jet prolongation of $\zeta'(\gamma)$ to $J^1L(X)\times_X J^1K$.

Theorem 2: *Locally, all Lagrangians V are characterized by a finite number of conditions.*

Proof: If

$$\gamma = f_i\frac{\partial}{\partial x_i} + g_M\tilde{A}_M\,,\quad \zeta'(\gamma) = f_i\frac{\partial}{\partial x_i} + \left(\frac{\partial g_N}{\partial x_i} - \Gamma_{kN}\frac{\partial f_k}{\partial x_i} + c^N_{AM}\Gamma_{iA}g_M\right)\frac{\partial}{\partial\Gamma_{iN}} - \frac{\partial f_k}{\partial x_i}e_k^{\ell}\frac{\partial}{\partial e_i^{\ell}}\,.$$

Thus, $L_{\overline{\zeta'(\gamma)}}(V\omega) = 0$, for any $\gamma\in\mathrm{aut}(G)$ if V is a solution of the following system of equations:

$$\frac{\partial V}{\partial B_{i\ell N}} + \frac{\partial V}{\partial B_{\ell iN}} = 0\,;\quad \frac{\partial V}{\partial H_i{}^{\ell}{}_h} + \frac{\partial V}{\partial H_h{}^{\ell}{}_i} = 0\,;\quad \frac{\partial V}{\partial x_i} = 0$$

$$\frac{\partial V}{\partial\Gamma_{iN}} + c^A_{MN}\Gamma_{\ell M}\frac{\partial V}{\partial B_{i\ell A}} = 0\,;\quad c^N_{AM}\left(\Gamma_{iA}\frac{\partial V}{\partial\Gamma_{iN}} + B_{i\ell A}\frac{\partial V}{\partial B_{i\ell N}}\right) = 0$$

$$\Gamma_{kN}\frac{\partial V}{\partial\Gamma_{\ell N}} + e_k^m\frac{\partial V}{\partial e_\ell^m} + (B_{kiN} - B_{ikN})\frac{\partial V}{\partial B_{\ell iN}} + (H^i_{kh} - H^i_{hk})\frac{\partial V}{\partial H_\ell{}^i{}_h} = 0\,.$$

1.5. Minimal coupling principle

Definition 4: The minimal coupling between a mater field and a gauge field, whose Lagrangian density are $L\omega$ and $V\omega$, is given by the variational problem $(\bar{\vartheta},(L + V)\omega)$. Its Euler-Lagrange equations are the interaction equations of both fields.

There exists no interaction modified Lagrangian

In fact, let V be a coordinate neighborhood on $\bar{Y}$ endowed with the coordinate system $(x_i, z_j, e_\ell{}^h, \Gamma_{kN}, p_{ij}, H_i{}^h{}_\ell, B_{ikN})$ defined in Section 1.2. And let $(x_i,\ldots,p'_{ij},\ldots)$ be a new coordinate system on V , where

$$p'_{ij} = e_i^{\;\ell} p_{\ell j} + \Gamma_{iN}\, a^N_{kj}\, z_k \quad .$$

Now, we can write $\bar{\vartheta}$ as:

$$\bar{\vartheta} = (dz_j - p'_{ij}dx_i) \circ \frac{\partial}{\partial z_j} + (de_h^{\;\ell} - H_i{}^h{}_\ell dx_i) \circ \frac{\partial}{\partial e_h^{\;\ell}} +$$

$$+ (d\Gamma_{kN} - B_{ikN}dx_i) \circ \frac{\partial}{\partial \Gamma_{kN}} \quad .$$

If the Lagrangian of the matter field was $L = L(z_j, p_{ij})$, in the former coordinate system, in the latter it is

$$L = L(z_j, e^i{}_\ell (p'_{ij} - \Gamma_{iN}\, a^N_{kj}\, z_k))$$

i.e., the *interaction modified Lagrangian* given by Kibble in [1].

2. THE POINCARE GAUGE FIELD (PGF) THEORY

2.1. A minimal coupling principle

Let H be a Lie group with a representation on some vector space $X \simeq \mathbf{R}^n$ and let G denote the *semidirect product* of H and X .

We can imbed the trivial H-bundle $p: G \longrightarrow X$ into $P = G \times X$:

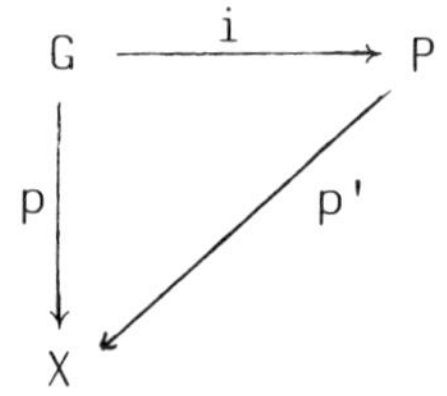

where $i(g) = (p(g),g)$. Obviously:

1) $(TP_G) = \text{aut } P = \text{gau } P \oplus \Gamma(TX)$.

2) Since P is trivial, $\text{gau } P \simeq C^\infty(X) \otimes_{\mathbf{R}} \mathfrak{g}$ and hence, an isomorphism j from $\text{aut } G$ into $\text{gau } P$ may be constructed as follows:

j (right G-invariant vector field on P associated with A) = right H-invariant vector field on G associated with A

for every element $A \in \mathcal{g}$.

Now let

a) $\pi_1 : E \longrightarrow X$ be a vector bundle associated with G with standard fibre F .

b) $\pi_2 : \mathrm{Aut}\ TX \longrightarrow X$ be the bundle whose fibre at any point $x \in X$ is:

$$(\mathrm{Aut}\ TX)_x = \{\text{group of linear automorphisms of } T_x(X)\} .$$

c) $\pi_3 : K \longrightarrow X$ be the bundle of connections on $p : G \longrightarrow X$.

Proposition 1: The fibred product $\mathrm{Aut}\ TX \times_X K$ is an *open subbundle of the bundle of connections on* $p' : P \longrightarrow X$ and we shall write $K' = \mathrm{Aut}\ TX \times_X K \xrightarrow{\pi'} X$.

Given a connection σ on $\pi_1 : E \longrightarrow X$ and a section Φ of the bundle $\mathrm{Aut}\ TX$, we can define the following structure form on $\bar{E}$:

$$\vartheta^{\sigma,\Phi}_{\bar{y}}(\bar{D}) = D_y - \sigma_x(D') + \psi \cdot \Phi^{-1}(D')$$

if $\bar{y} = (y,\psi) \in \bar{E}$. Therefore:

Proposition 2: The set of structure forms on $\bar{E}$ can be parametrized by cross sections of K' .

In terms of the coordinate system on $\bar{E}$ (x_i, z_j, p_{ij}) the form $\vartheta^{\sigma,\Phi}$ is:

$$\vartheta^{\sigma,\Phi} = [dz_j - (e_i^{\ell}(x) p_{\ell j} + \Gamma_{ih}(x) a^h_{kj} z_k) dx_i] \circ \frac{\partial}{\partial z_j}$$

if $e_i^{\ell}(x) \dfrac{\partial}{\partial x_\ell} = \Phi^{-1}(\dfrac{\partial}{\partial x_i})$; compare with Eq. (1) of Section 3.1.1.

We can also define on $\bar{Y} = \bar{E} \times_X J^1 \mathrm{Aut}\ TX \times_X J^1 K$ a structure form $\bar{\vartheta}$:

$$\bar{\vartheta}_{\bar{z}}(\bar{D}) = \vartheta^{\sigma,\Phi}_{\bar{y}}(\bar{D}_1) + \vartheta'_{j^1_x \Phi}(\bar{D}_2) + \vartheta''_{j^1_x \sigma}(\bar{D}_3)$$

where $\bar{z} = (\bar{y}, j^1_x \Phi, j^1_x \sigma)$ and ϑ' , ϑ'' are the structure forms canonically defined on $J^1 \mathrm{Aut}\ TX$ and $J^1 K$, respectively.

Let us assume an auxiliary symmetric and nondegenerate, metric T_2 to be given in the manifold X , and let ω be the volume n-form canonically associated to T_2 .

Definition 1: We shall call the Lagrangian density of a matter field each form $L\omega$, defined on $E \times_X J^1 \mathrm{Aut}\, TX$, where L is an arbitrary function on $\bar{E}$ and ω the n-form on Aut TX :

$$\omega_{(x,\Phi_x)}(D_1,\ldots,D_n) = \omega_x(\Phi_x^{-1}(\pi_{2*}D_1),\ldots,\Phi_x^{-1}(\pi_{2*}D_n))$$

Remark: When G is the Poincaré group and H the Lorentz group, T_2 is the Minkowskian metric and $\omega_{(x,\Phi_x)}$ is the volume form associated to T_2 and written in "an arbitrary system of coordinates" on $\mathbf{R}^4$.

The same as in the C-K-S theory, expounded above, we may define on $\bar{Y}$ a coordinate system $(x_i,z_j,e_\ell{}^h,\Gamma_{kN},p_{ij},H_i{}^h{}_\ell,B_{ikN})$ — now $(e_\ell{}^h(x,\Phi_x))$ is the matrix of the automorphism Φ_x^{-1} with respect to the basis $(\frac{\partial}{\partial x_1})_x,\ldots,(\frac{\partial}{\partial x_n})_x$ — such that, with respect to it, the expression in coordinates of $\bar{\vartheta}$ coincides with equation (B) of Section 1.2 . By an identical change of coordinates, we pass from the interaction-free Lagrangian to the modified Lagrangian given by Helh et al. in [3] .

2.2. Lagrangian density of the matter field

2.1.1. Let us denote by Y the fibred product $E \times_X K'$. *We shall construct a morphism of $C^\infty(X)$-modules* ζ:aut P $\longrightarrow \Gamma(TY)$:

a) $\pi_1:E \longrightarrow X$ is a vector bundle associated with G . Thus, there is a representation ξ:aut G $\longrightarrow \Gamma(T_{\pi_1}E)$ such that,

$$\xi(\tilde{A}_N) = a^N_{hk}\, z_h \frac{\partial}{\partial z_k} \quad ; \quad \xi(\tilde{B}_j) = g_{ij}(x)\frac{\partial}{\partial x_i}$$

where $\{A_N\}$ is a basis of the Lie algebra H , $\{A_N,B_j\}$ is a basis of g and $\{\tilde{A}_N,\tilde{B}_j\}$ are the corresponding H-invariant vector fields on G .

Using ξ , we define the following morphism of $C^\infty(X)$-modules ζ_1:aut G $\longrightarrow \Gamma(TY)$:

Let us assume that $D\varepsilon$aut G and let $D = D^H + D^F$ be its decomposition in vertical and horizontal components, respectively, according to the flat connection on $p:G \longrightarrow X$,

$$\zeta_1(D)_y = -\xi(D^H)_{(x,e)} + \xi\sigma(\Phi^{-1}(p_* D^F)_x)$$

where $y = (x,e,\Phi,\sigma)$ is a point of Y given by a vector $e\varepsilon F$, an automorphism Φ of $T_x(X)$ and a connection σ .

The expression in coordinates of ζ_1 is:

$$\zeta_1(\tilde{A}_N) = -a^N_{hk}\, z_h \frac{\partial}{\partial z_k}$$

$$\zeta_1(\tilde{B}_j) = e^i{}_\ell\, g_{\ell j}(\frac{\partial}{\partial x_i} + \Gamma_{iN}\, a^N_{hk}\, z_h \frac{\partial}{\partial z_k})$$

b) $K' \xrightarrow{\pi'} X$ is an open subbundle of the bundle of connections on P and thus there is another representation (see Section 1.3.2) $\xi' : \text{Aut } P \longrightarrow \Gamma(T_{\pi'}K')$. Let us denote by p_2 the canonical projection of Y on K' . All vector fields on K' produce by the pull-back p_2^* a field on Y . Composing ξ' with p_2^* gives rise to a morphism of $C^\infty(X)$-modules $\zeta_2 : \text{aut } P \longrightarrow \Gamma(TY)$.

c) Finally, if we continue to denote by ζ_1 the morphism of aut P into $\Gamma(TY)$ which is the result of composing the projection of aut P onto gau P with $\zeta_1 \circ j$, we may define:

$$\zeta = \zeta_1 - \zeta_2 : \text{aut } P \longrightarrow \Gamma(TY) \ .$$

Definition 2: A matter field on $\bar{Y}$, with Lagrangian density $\mathcal{L}\omega$, is aut P-invariant if $L_{\overline{\zeta(\gamma)}}(\mathcal{L}\omega) = 0$ for all $\gamma \in \text{aut } P$, where $\overline{\zeta(\gamma)}$ is the ϑ-lifting of $\zeta(\gamma)$ to $\bar{Y}$.

2.2.2. *The morphism* ζ *allows us to construct a representation, which we shall also denote by* ζ *, of the Lie algebra* g *into* $\Gamma(TY)$:

$$\zeta : A \longrightarrow \Gamma(RA) \ ,$$

where RA is the right G-invariant vector field on P associated with A .

In particular,

$$\zeta(A_N) = -a^N_{hk} z_h \frac{\partial}{\partial z_k} - c^\ell_{hN} e_i{}^h \frac{\partial}{\partial e_i{}^\ell} - (c^M_{QN} \Gamma_{iQ} + c^M_{kN} e_i{}^k) \frac{\partial}{\partial \Gamma_{iM}}$$

$$\zeta(B_j) = e^i{}_\ell g_{\ell j} (\frac{\partial}{\partial x_i} + \Gamma_{iN} a^N_{hk} z_h \frac{\partial}{\partial z_k}) - (c^\ell_{Mj} \Gamma_{iM} + c^\ell_{hj} e_i{}^h) \frac{\partial}{\partial e_i{}^\ell} -$$

$$- (c^N_{Mj} \Gamma_{iM} + c^N_{hj} e_i{}^h) \frac{\partial}{\partial \Gamma_{iN}}$$

where the $c^\ell_{kN}, c^M_{QN}, \ldots$ are the structure constants of g (for example, $[B_h, A_N] = c^j_{hN} B_j + c^M_{hN} A_M$) .

Let us finally define a second representation $\zeta' : g \longrightarrow \Gamma(TY)$ in the following way:

$$\zeta' : A \longrightarrow \sigma_0 \pi_{1*} \xi(\tilde{A})$$

where σ_0 is the flat connection on F . We shall now have,

$$\zeta'(A_N) = 0 \quad ; \quad \zeta'(B_j) = g_{ij} \frac{\partial}{\partial x_i} \ .$$

Definition 3: A matter field $(Y, L\omega)$ is said to the $\mathcal{g}$-invariant if, for all $A \varepsilon \mathcal{g}$, we have simultaneously:

$$L_{\overline{\zeta(A)}}(L\omega) = 0 \quad ; \quad L_{\overline{\zeta'(A)}}(L\omega) = 0 \quad ,$$

$\overline{\zeta(A)}$ and $\overline{\zeta'(A)}$ being the $\bar{\vartheta}$-liftings of $\zeta(A)$, $\zeta'(A)$ to $\bar{Y}$.

Theorem 1 (see [13]): *A matter field* $(\bar{Y}, L\omega)$ *is* aut P-*invariant if and only if it is* $\mathcal{g}$-*invariant.*

Proof: Let us assume that $(\bar{Y}, L\omega)$ is $\mathcal{g}$-invariant. To show that it is also aut P-invariant it is only necessary to prove that $L_{\overline{\zeta(\gamma)}}(L\omega) = 0$ when $\gamma = a_i \frac{\partial}{\partial x_i}$ and when $\gamma = f_j RB_j + f_N RA_N$.

a) If $\gamma = a_i \frac{\partial}{\partial x_i}$, we have,

$$\zeta(\gamma) = a_i \frac{\partial}{\partial x_i} - \Gamma_{kN} \frac{\partial a_k}{\partial x_i} \frac{\partial}{\partial \Gamma_{iN}} - e_k^{\ell} \frac{\partial a_k}{\partial x_i} \frac{\partial}{\partial e_i^{\ell}} \quad \text{and}$$

$$L_{\overline{\zeta(\gamma)}}(L\omega) = a_i \, L_{\overline{\zeta(\frac{\partial}{\partial x_i})}}(L\omega) \ .$$

Since $\zeta'(B_j) = g_{ij} \frac{\partial}{\partial x_i}$, where $|g_{ij}(x)| \neq 0$, condition $L_{\overline{\zeta(\frac{\partial}{\partial x_i})}}(L\omega) = 0$, for $i = 1, \ldots, n$, is equivalent to $L_{\overline{\zeta'(B_j)}}(L\omega) = 0$, for $j = 1, \ldots, n$. Thus, $\mathcal{g}$-invariance of $L\omega \Rightarrow L_{\overline{\zeta(\gamma)}}(L\omega) = 0$.

b) If $\gamma = f_j RB_j + f_N RA_N$, we shall have:

$$\zeta(\gamma) = f_j \zeta(B_j) + f_N \zeta(A_N) - \frac{\partial f_N}{\partial x_i} \frac{\partial}{\partial \Gamma_{iN}} - g_{\ell r} \frac{\partial f_r}{\partial x_i} \frac{\partial}{\partial e_i^{\ell}} \quad \text{and}$$

$$\overline{\zeta(\gamma)}(p_{h\ell}) = (f_j \overline{\zeta(B_j)} + f_N \overline{\zeta(A_N)})(p_{h\ell})$$

from where it may be deduced that

$$L_{\overline{\zeta(\gamma)}}(L\omega) = f_j L_{\overline{\zeta(B_j)}}(L\omega) + f_N L_{\overline{\zeta(A_N)}}(L\omega) = 0$$

if, as we assume, $L = L(x_i, z_j, p_{h\ell})$ and $L\omega$ is $\mathcal{g}$-invariant.
The opposite is shown in the same way.

2.3. Characterization of all possible Lagrangians V of the generalized gauge field K'.

Definition 4: A generalized gauge field is a classical field defined on K' whose Lagrangian density $L\omega$ in aut P-invariant.

Let us finish the chapter with a final

Remark: The main differences between the C-K-S and PGF theories may be summarized as follows:

1) In the former, we start from a single principal H-bundle G. The matter fields are defined on vector bundles associated with G and the gauge fields on the fibred product $L(X) \times_X K$. In the PGF theory, *two* principal bundles $G \hookrightarrow P$ are considered (*both trivial!*). The matter fields are defined on vector bundles associated with G and the gauge fields on an open subbundle of the *bundle of connections on* P.

2) The transformation properties of the gauge potentials $(\Gamma_{k\ell}, e_j^{\ell})$ are given in the C-K-S theory by the canonical representation of aut G into $L(X) \times_X K$ (see Section 1.4). In the second theory the canonical representation of aut P into K' is used to obtain such transformations properties. Thus:

3) From a physical point of view, both theories are differentiated in that the Lagrangians V of the corresponding generalized gauge fields should fulfil different local conditions. Compare Section 1.4 with equations given in [14].

REFERENCES

[1] T.W.B. Kibble - *Lorentz Invariance and Gravitational Field* - J. Mathematical Phys., 2 (1961), pp. 212-221.

[2] D.W. Sciama - In *Recent Developments in General Relativity* - Pergamon Press (1962), pp. 415-440.

[3] F.W. Held, P.v.d. Heyde, G.D. Kerlick and J.M. Nester - *General Relativity with spin and torsion: Foundations and Prospects* - Rev. Modern. Phys., 48 (1976), pp. 393-416.

[4] P. Von der Heyde - *The field equations of the Poincaré gauge theory of gravitation* - Phys. Lett. 58A, (1976), pp. 141-143.

[5] F.W. Hehl, J. Nitsch and P.v.d. Heyde - *Gravitation and the Poincaré gauge field theory with quadratic Lagrangian* - In *General Relativity and Gravitation* - Vol. 1, Plenum Publishing Corporation (1980), pp. 329-355.

[6] Y. Ne'eman and T. Regge - *Gauge theory of gravity and supergravity on a group manifold* - Rivista del Nuovo Cim., 1, nº 5 (1978).

[7] Y. Ne'eman - *Gravity, groups and gauges* - In *General Relativity and Gravitation* - Vol. 1, Plenum Publishing Corporation (1980), pp. 309-327.

[8] A. Trautman - *Fiber bundles, gauge fields and gravitation* - *In General Relativity and Gravitation* - Vol. 1, Plenum Publishing Corporation (1980), pp. 287-308.

[9] S. Kobayashi and K. Nomizu - *Foundations of Differential Geometry* - Vol. 1, Interscience (1963).

[10] S. Sternberg - *On the role of field theories in our physical conception of geometry* - In *Differential Geometrical Methods in Mathematical Physics* - Bonn, 1977. Lecture Notes in Mathematics, Vol. 676, pp. 1-80.

[11] A. Pérez-Rendón - *Gravitation and Gauge Theories* - Prepint, Universitá degli Studi de Firenze.

[12] P.L. García - *Gauges Algebras, Curvature and Symplectic Structure* - J. Differential Geometry, 12 (1977), pp. 209-227.

[13] A. Pérez-Rendón - *The Yang-Mills trick and gravitational interactions* - In *Symplectic Geometry* - Research Notes in Math. nº 80, Pitman (1983), pp. 217-226.

[14] - *Lagrangiennes dans les théories jauge par rapport au groupe de Poincaré* - To be published in Rendiconti del Seminario Matematico. Torino.

GAUGE INDEPENDENT SYMMETRIES & WAVEFUNCTIONS FOR SYSTEMS OF IDENTICAL PARTICLES

J.M. Selig

D.A.M.T.P., The University, Liverpool L69 3BX, U.K.

Abstract

We demonstrate a correspondence, due to Atiyah, between G-bundles over a G-space M and vector bundles over the quotient M/G. We use this to investigate the actions of S_3 on the trivial associated SU(2) bundle over the configuration space of three distinguishable particles moving in $\mathbb{R}^3$, $\tilde{C}_3(\mathbb{R}^3)$. These correspond to the possible SU(2) bundles over $C_3(\mathbb{R}^3)$, the configuration space of three identical particles moving in $\mathbb{R}^3$.

1. INTRODUCTION

Souriau [8] has shown than only two inequivalent prequantizations exist for a system of identical elementary particles. The existence of two non-isomorphic complex hermitian line bundles over the phase space of a system of n identical particles is taken to define the boson-fermion superselection rule. Let:-

$$\tilde{C}_n(M) = (\underbrace{M \times \cdots\cdots \times M}_{n \text{ factors}}) \setminus \{\text{diagonals}\}$$

be the configuration space of n distinguishable particles moving in M. Then the configuration space of n identical particles moving in M is the quotient:-

$$C_n(M) = \tilde{C}_n(M)/S_n$$

where the Symmetric group S_n acts by permuting the particles, see [3]. Now if $M = \mathbb{R}^m$ $(m > 2)$ there are again two isomorphism classes of complex line bundles over $C_n(\mathbb{R}^m)$, corresponding directly with Souriau's result. Further, it has been shown

S. Sternberg (ed.), Differential Geometric Methods in Mathematical Physics, 75–80.

[4,6] that sections of the trivial bundle pull-back to symmetric sections of the trivial bundle over $\tilde{C}_n(\mathbb{R}^m)$. While sections of the essential bundle pull-back to antisymmetric sections of the trivial bundle over $\tilde{C}_3(\mathbb{R}^m)$. Hence the interpretation in terms of bosons and fermions. However this nice correspondence depends on a particular choice of trivialization of the trivial bundle over $\tilde{C}_n(\mathbb{R}^m)$. In general a gauge transformation (a bundle automorphism which projects to the identity on the base) will destroy the symmetry.

We will show how the symmetry can be recovered by allowing the action of S_n on a bundle over $\tilde{C}_n(\mathbb{R}^m)$ to be by bundle automorphism not just representations in the fibres. Gauge transformations then change one such action into another. A theorem of Atiyah [1] gives us a 1-to-1 correspondence between equivalence classes of actions (under gauge transformation) and isomorphism classes of bundles over $C_n(\mathbb{R}^m)$.

In the last section we give an example. We investigate the possible symmetries of wavefunctions for a system of three identical particles moving in $\mathbb{R}^3$, where wavefunctions are considered as sections of some hermitian vector bundle.

2. THE GENERAL CASE

Consider the following situation:-

$$\begin{array}{ccc} \beta^*E & \longrightarrow & E \\ \downarrow & & \downarrow p \\ \tilde{B} & \xrightarrow{\ \beta\ } & B \end{array}$$

where $\beta:\tilde{B} \to B$ is a regular finite covering, that is $\Pi_1(\tilde{B}) \trianglelefteq \Pi_1(B)$; and $\xi = (E, p, B)$ is a vector bundle. The pull-back of E by β, β^*E is given by:-

$$\beta^*E = \{(\tilde{b}, v) \in \tilde{B} \times E \mid \beta(\tilde{b}) = p(v)\}$$

There is a natural action of $G = \Pi_1(B)/\Pi_1(\tilde{B})$ on β^*E:-

$$\Phi : G \times \beta^*E \to \beta^*E$$

$$\Phi(g)(\tilde{b},v) = (g.\tilde{b},v) \quad , \quad \forall g \in G,\ (\tilde{b},v) \in \beta^*E$$

where g.b denotes the image of $\tilde{b}$ under the covering transformation labelled by $g \in G$.

In a bundle $\tilde{\xi} = (\tilde{E}, \tilde{p}, \tilde{B})$ isomorphic to $(\beta^*E, \beta^*p, \tilde{B})$ we have an induced action:-

$$(id_G \times \tau^{-1}) \circ \Phi \circ (id_G \times \tau): \ G \times \tilde{E} \to \tilde{E}$$

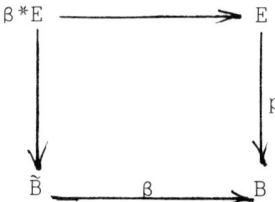

where $\tau: \tilde{E} \to \beta^*E$ is an isomorphism, in other words a gauge transformation.

Conversely given an action Ψ of G on a bundle $\tilde{\xi} = (\tilde{E}, \tilde{p}, \tilde{B})$ which satisfies:-

$$\Psi : G \to \mathrm{Aut}\,(\tilde{\xi})$$

$$\tilde{p} \circ \Psi(g)(\tilde{e}) = g.\tilde{p}(\tilde{e}), \ \forall \tilde{e} \in \tilde{E}$$

then $\tilde{E}/\Psi$ is a vector bundle over $\tilde{B}/G \cong B$. The second condition is just to ensure that Ψ projects to the covering transformations on $\tilde{B}$.

Since the action of G on $\tilde{B}$ is free Atiyah's theorem [1] tells us that there is a 1-to-1 correspondence between the actions of G defined above, modulo gauge equivalence and isomorphism classes of vector bundle over B. This is because gauge equivalence of actions, defined by:-

$$\Phi \sim \Phi' \Leftrightarrow \exists \tau: \tilde{E} \to \tilde{E}' \text{ such that } \Phi = (id_G \times \tau^{-1}) \circ \Phi \circ (id_G \times \tau)$$

is clearly equivalent to a G-isomorphism of the G-bundles $(\tilde{E}, \tilde{p}, \tilde{B}, \Phi)$ and $(\tilde{E}', \tilde{p}', \tilde{B}, \Phi')$.

A section $s : \tilde{B} \to \tilde{E}$ of a G-bundle $\tilde{\xi} = (\tilde{E}, \tilde{p}, \tilde{B}, \Phi)$ is equivariant if and only if it satisfies

$$S(\tilde{b}) = \Phi(g)(S(g^{-1}.\tilde{b})), \ \forall \tilde{b} \in \tilde{B}, \quad g \in G$$

Under the quotient map $p : \tilde{\xi} \to \tilde{\xi}/G \cong \xi$, an equivariant section of G will be mapped to a section of ξ. Conversely the pull-back by p of a section of ξ is an equivariant section of $\beta^*\xi$ with the natural action of G. So we can identify the vector spaces of sections:-

$$\Gamma(\xi) \cong \Gamma(\beta^*\xi)^G$$

3. WAVEFUNCTIONS FOR THREE IDENTICAL PARTICLES MOVING IN $\mathbb{R}^3$

It is known that vector bundles over $C_3(\mathbb{R}^3)$, with structure group SU(2), are classified by their second Chern class $c_2 \in H^4(C_3(\mathbb{R}^3), \mathbb{Z}) \cong \mathbb{Z}_3$; [5]. Since the cohomology of the universal cover; $\tilde{C}_3(\mathbb{R}^3)$ is all free, these three bundles all pull-back to the trivial bundle over $\tilde{C}_3(\mathbb{R}^3)$. Hence on $\tilde{C}_3(\mathbb{R}^3) \times \mathbb{C}^2$, in particular, we should be able to find three inequivalent actions of $\Pi_1(C_3(\mathbb{R}^3)) \cong S_3$. To find these we first recall some facts about the automorphism group of a trivial bundle [7, 9].

For any trivial vector bundle $(M \times V, P_1, M)$ with structure group H any automorphism can be written as a pair:-

$$(t, \sigma), \ t \in \mathrm{Diff}(M), \sigma \in \mathrm{Map}(M, H)$$

The image of a point $(m,u) \in M \times V$ under such an automorphism is given by:-

$$(t, \sigma)(m, u) = (t(m), \sigma(t(m)u)$$

Hence composition of automorphisms gives $\mathrm{Aut}(M \times V, p_1, M)$ the structure of a semi-direct product of Diff(M) with Map(M, H). A gauge transformation has the form:-

$$(id_M, \sigma), \sigma \in \mathrm{Map}(M, H).$$

First we investigate the representations, that is an action of the form:-

$$\Phi(g) = (tg, \sigma_D(g)) \qquad , \forall g \in G$$

where tg is a covering transformation labelled by g and $\sigma_D(g)$ is the constant map:-

$$\sigma_D(g) : M \to D(g),$$

for some representation of G in H.
In our case $G \equiv S_3 = \langle x, y \mid x^2 = y^3 = e, xy = y^2x\rangle$ and $H \equiv SU(2)$.
Only two reps are possible:-

$$D_0(x) = I \quad , \quad D_0(y) = I$$

and

$$D_1(x) = -I \quad , \quad D_1(y) = I$$

Now these representations have Chern classes [9] since they induce maps between the universal classifying spaces

$$D_* : BS_3 \to BSU(2)$$

A simple calculation gives that:-

$$(\tilde{C}_3(\mathbb{R}^3) \times \mathbb{C}^2)/D_0 \cong (\tilde{C}_3(\mathbb{R}^3) \times \mathbb{C}^2)/D_1 \cong C_3(\mathbb{R}^3) \times \mathbb{C}^2$$

That is; the quotients by both these actions give the trivial bundle over $C_3(\mathbb{R}^3)$. So there must be a gauge transformation which changes D_0 into D_1, that is:-

$$(id, \chi^{-1}) \circ (tg, \sigma_{D_0}(y)) \circ (id, \chi) = (tg, \sigma_{D_1}(g)), \forall g \in S_3$$

which simplifies to:-

$$\chi^{-1}(g.\tilde{b})D_0(g)\chi(\tilde{b}) = D_1(g) \quad , \quad \forall \tilde{b} \in \tilde{C}_3(\mathbb{R}^3), g \in S_3$$

This is possible if $\chi \in \mathrm{Map}(\tilde{C}_3(\mathbb{R}^3), SU(2))$ is antisymmetric. To construct χ think of point in $\tilde{C}_3(\mathbb{R}^3)$ as triangles with vertices at $(\underline{r}_1, \underline{r}_2, \underline{r}_3)$, $\underline{r}_i \in \mathbb{R}^3$, $i \neq j$ $\underline{r}_i \neq \underline{r}_j$. Consider the two antisymmetric vector valued function:-

$$\underline{A}=\underline{r}_1\times\underline{r}_2+\underline{r}_2\times\underline{r}_3+\underline{r}_3\times\underline{r}_1$$

$$\underline{B}=(\underline{r}_2-\underline{r}_3)[(\underline{r}_3-\underline{r}_1).(\underline{r}_1-\underline{r}_2)]+(\underline{r}_3-\underline{r}_1)[(\underline{r}_1-\underline{r}_2).(\underline{r}_2-\underline{r}_3)] +(\underline{r}_1-\underline{r}_2)[(\underline{r}_2-\underline{r}_3).(\underline{r}_3-\underline{r}_1)]$$

$\underline{A}=0$ if and only if the triangle $(\underline{r}_1, \underline{r}_2, \underline{r}_3)$ is colinear and $\underline{B}=0$ if and only if the triangle is equilateral. Also $\underline{A}.\underline{B}=0$ always, so $\underline{M}=\underline{A}+\underline{B}$ is a nowhere zero antisymmetric vector valued function. The required SU(2) valued function χ is thus:-

$$\chi(\underline{r}_1, \underline{r}_2, \underline{r}_3) = \frac{-i}{|\underline{M}|}\underline{M}.\underline{\sigma}$$

where $\underline{\sigma} = (\sigma_x, \sigma_y, \sigma_z)$, the Pauli matrices.

We return to the problem of finding the actions corresponding to the non-trivial bundles over $C_3(\mathbb{R}^3)$. Let $\hat{C}_3(\mathbb{R}^3)$ be the two-fold regular cover of $C_3(\mathbb{R}^3)$, so we have:-

$$\tilde{C}_3(\mathbb{R}^3) \xrightarrow{\alpha} \hat{C}_3(\mathbb{R}^3) \xrightarrow{\gamma} C_3(\mathbb{R}^3)$$

and $\gamma \circ \alpha = \beta$. Now $\Pi_1(\hat{C}_3(\mathbb{R}^3)) \cong \mathbb{Z}_3$ and

$$H^*(\hat{C}_3(\mathbb{R}^3), \mathbb{Z}) \cong \mathbb{Z}[\omega, \nu]/\{\omega^3, 3\omega, \nu^2, \omega\nu\} \quad \deg\omega = \deg\nu = 2$$

also $\gamma^* : H^4(C_3(\mathbb{R}^3), \mathbb{Z}) \to H^4(\hat{C}_3(\mathbb{R}^3), \mathbb{Z})$ is an isomorphism. The fundamental 1-dimensional rep, R of $\mathbb{Z}_3$ has total Chern class $(1+\omega)$, so the rep $R + \bar{R}$ has total Chern class $(1+2\omega^2)$. This is in the image of γ^*, hence the action for the bundle over $\tilde{C}_3(\mathbb{R}^3)$ with $c_2 = 2 \bmod 3$ has:-

$$\Phi(y) = (ty, -1/2 - i\sqrt{3}/2\sigma_z)$$

The action of $S_3/\mathbb{Z}_3 \cong \mathbb{Z}_2$ on this bundle can be pulled-back to $\tilde{C}_3(\mathbb{R}^3)$ to give a function with the following symmetry property.

$$\Phi(x) = (tx, k + i\ell\sigma_z + im\sigma_x + im\sigma_y)$$

with $k^2+l^2+m^2+n^2 = 1$ and the pair of functions $\begin{pmatrix} k(\underline{r}_1, \underline{r}_2, \underline{r}_3) \\ l(\underline{r}_1, \underline{r}_2, \underline{r}_3) \end{pmatrix}$ transform under particle interchange with the 2-dimensional irrep of S_3 while $m(\underline{r}_1, \underline{r}_2, \underline{r}_3)$ and $n(\underline{r}_1, \underline{r}_2, \underline{r}_3)$ are totally antisymmetric. We will not discuss the action for the other non-trivial bundle.

Acknowledgments

I wish to thank F.J. Bloore for guidance and supervision and also S.E.R.C. for financial support.

References

1. Atiyah, M.F. "K-Theory" 1967 Amsterdam: Benjamin inc pp.35-41
2. Atiyah, M.F. I.H.E.S. Publications Mathematiques 1961 9 pp.23-64.
3. Bloore, F.J. "Lecture Notes in Mathematics vol. 836" 1980 Berlin: Springer-Verlag pp.1-6.
4. Bloore, F.J. and Swarbrick S.J. J.Math.Phys. 1978 19 pp. 878-879.
5. Bloore, F.J., Bratley, I. and Selig J.M. J. Phys. A 1983 16 pp.729-736.
6. Doebner, H.D., Stovicek, P. and Tolar, J. I.C.T.P. Trieste preprint 1981 IC/81/116.
7. Husemoller, D. " Fibre Bundles" 1966 New York: McGraw-Hill p.59.
8. Souriau, J.M. "Structure des systemes dynamiques 1970 Paris: Dunod pp.363-392.
9. Trautman, A. "Lecture Notes in Physics vol.129" Berlin: Springer-Verlag pp.114-120.

SUPERSPACES AND SUPERMANIFOLDS

J. Hoyos (*)(†), M. Quirós (†), F.J. de Urries(*,*)(†)
J. Ramírez Mittelbrunn (*,*) (†)

(*) Dpto. de Matemáticas, (*,*) Dpto. de Física.
Universidad de Alcalá de Henares. (Madrid) Spain.
(†) Instituto de Estructura de la Materia del C.S.I.C.
Serrano, 119. Madrid (6) Spain.

ABSTRACT

Generalized superspaces are constructed using as basic building block the concept of Grassmann-Banach algebra. The category of B-spaces, wider than that of superspaces, is introduced to discuss B-linear and multi-B linear maps between superspaces. The analysis in generalized superspaces is studied. G-differentiability of functions is defined, and the properties of G^k functions studied with special attention to extension theorem and canonical expansion (superfield expansion). Supermanifolds are introduced as Banach manifolds modeled on a superspace. The body (real part) of supermanifolds is analyzed arriving at the concept of ρ-supermanifolds. Tangent, contangent and tensor bundles of ρ-supermanifolds are discussed. The elements of the theory of Lie supergroups and principal superfiber bundles as ρ-supermanifolds are outlined.

1. INTRODUCTION

In recent years there has been a lot of interest from physics and mathematics in the study of superspaces and supermanifolds. Naively these concepts correspond to the extension of the vector space and manifold notions respectively to the case of having both commuting and anticommuting coordinates. There are two approaches to a formalized concept of supermanifold. In Berezin-Leites (1) and Kostant (2) approach a supermanifold is given by an ordinary C^∞ manifold X furnished with a ring of functions $A[X]$ enriched with anticommuting elements.

S. Sternberg (ed.), Differential Geometric Methods in Mathematical Physics, 81–99.

$A[X] \approx C^\infty[X] \otimes \Lambda q$, where Λq is the exterior algebra expanded by q generators. A specific model for the algebra $A[X]$ as the space of cross sections of a bundle with base X and fiber Λq has been given by Dell-Smolin (3). So such things as "points" of superspace having both commuting and anticommuting coordinates do not exist in this approach.

The second approach to superspaces and supermanifolds is due to Rogers (4). Here superspaces are introduced as products of even and odd parts of a Grassman algebra. Differentiation of functions with domain in a superspace is done with respect to the linearity associated to the Grassmann product (B-linearity). The differentiation notion so obtained is called G-differentiation. Accordingly G^∞ or supersmouth supermanifolds are Banach manifolds locally G-diffeomorphic to a superspace.

In this lecture we present a theory of superspaces and supermanifolds generalizing Rogers's approach (5). The aim is to give a mathematical model capable to be used as a framework to formulate and solve geometrical problems in supersymmetric theories. As we shall see the main ideas of analysis and differential geometry can be translated naturally to the corresponding "supernotions" in this approach. It has also the advantage of admiting supermanifolds with non-trivial topology in the anticommuting sector. On the other hand there is room for infinite dimensional Grassmann algebra, that avoids undesirable vanishings of Green's functions when dealing with field theory. Finally Berezin-Leites-Kostant manifolds can be realized as G^∞-supermanifolds.

We start the discussion introducing the basic building block of the theory, namely the Banach-Grassmann algebra which will replace the real numbers of ordinary analysis and differential geometry.

2. THE BANACH-GRASSMANN ALGEBRA

Let N be the set of natural numbers, and F the set of parts of N. For each $\emptyset \neq M \subset F$, when we write $M = \{i_1, i_2, \ldots, i_k\}$ we assume $i_1 < i_2 < \ldots < i_k$.

Definition 2.1 A Banach-Grassmann algebra B is a Banach algebra B with unity such that there exist a denumerable subset $\{\beta_i\}_{i=1}^{\infty}$ which, i) $\beta_i \beta_j = -\beta_j \beta_i \ \forall\, i,j \in N$. ii) if $\beta_\phi = 1$ and $\beta_M = \beta_{i_1} \ldots \beta_{i_k}$ for $M = \{i_1, \ldots, i_k\}$, then $\{\beta_M\}$ is a Banach space basis for B. iii) $\forall$ $H \subset F$ and $a = \sum_{M \in F} a_M \beta_M$, $p_H(a) = \sum_{M \in H} a_M \beta_M$ is a linear map from B to B.

We call to p_H projections, β-projections. The continuity of β-projections is not imposed directly in Definition 2.1 but it is a consequence of it.

Proposition 2.2 For each $H \subset F, p_H: B \to B$ is a continuous linear projection. Furthermore $B = p_H(B) \oplus p_{H'}(B)$, $H' = F-H$ and $p_{H'} = I-p_H$, I is the identity on B.

Taking $H = \{M \in F \mid \text{card}(M) \text{ is even}\}$ and calling $B^0 = p_H(B)$, $B^1 = p_{H'}(B)$ we have $B = B^0 \oplus B^1$. It is easy to see that $1 \in B^0$, $\beta_i \in B^1$, $i = 1,2,\ldots$, $B^r B^s \subset B^{r+s}$ for $r,s \in Z_2 = \{0,1\}$ and $ab = (-1)^{rs}\, ba$ for $a \in B^r$, $b \in B^s$. So B is a commutative Z_2-graded algebra.

Another important decomposition of B is given by $B = R \oplus B'$ that we obtain taking $H = \{\emptyset\}$. We call $r = p_H$ the body map and $s = I-r$ the soul map of B. We remark that r is an algebra homomorphism. On the other hand B' is the closure of the set B'' of nilpotent elements of B. So $B^1 \subset B'' \subset \overline{B''} = B'$.

Proposition 2.3 i) For $a \in B'$ and $0 < \theta < 1$, $\exists\, \lambda \geq 0$ such that $\forall n \geq 0$, $\|a^n\| \leq \lambda\, \theta^n$ ii) $a \in B$ is invertible iff $r(a) \neq 0$. Furthermore

$$a^{-1} = r(a)^{-1} \sum_{n=0}^{\infty} (-1)^n \left(s(a)/_{r(a)}\right)^n \tag{1}$$

3. LINEAR ALGEBRA IN B-SPACES

We discuss now linear algebra in B-spaces. B-spaces include B-modules and superspaces as relevant particular cases. Accordingly we divide the study of linear algebra in three parts: B-modules, general B-spaces and superspaces.

3. A. B-Modules

A very complete discussion of linear algebra on B-modules is given in reference (1). In what follows we shall present only some points relevant to our construction and the reader is referred to (1) for further details.

Let B be a Grassmann-Banach algebra.

Definition 3.1 A Banach Z_2 graded left B-module is a Banach B-space V that is also a left B-module and can be decomposed as a direct sum of Banach subspaces $V = V^0 \oplus V^1$, with $B^r V^s \subset V^{r+s}$. In addition $\|av\| \leq \|a\|\,\|v\|$, for $a \in B$, $v \in V$. V is free if it has a basis of homogeneous elements. In what follows all B-modules that will be considered are free unless other thing is specified. On the other hand, since B is graded commutative, each right B-module can be considered as a left

B-module, and conversely, in the following form:

$$a_r v_s = (-1)^{rs} v_s a_r, \quad a_r \in B^r, \quad v_s \in V^s \tag{2}$$

Definition 3.2 The body map associated to a fixed basis $\{e_i\}_{i=1}^{m+n}$ of V is the map

$$R\left(\sum_{i=1}^{m+n} a^i e^i\right) = (r(a^1), \ldots, r(a^{m+n})) \tag{3}$$

R is linear, continuous, and $R(av) = r(a)R(v)$ for $a \in B$, $v \in V$

Next we introduce B-linear maps between B-modules.

Defintition 3.3 A B-linear map from V into V' is a continuous linear map $T \in L(V,V')$ such that $(av)T = a(vT)$, $a \in B$, $v \in V$. The set of B-linear maps from V into V' will be denoted by $L_B(V,V')$.

We give to $L_B(V,V')$ a B-module structure. The scalar multiplication is given by $u(Ta) = (uT)a$. The degree is determined by $T = r$ iff $V^s T \subset V'^{r+s}$ $r,s = 0,1$. $L_B(V,V')$ is a closed subspace of $L(V,V')$ and so it is a Banach space. The dimension of $L_B(V,V')$ as free B-module is $(mm'+nn', mn'+m'n)$. As a B-module $L_B(V,V')$ has a body map R defined on it. Given $T \in L_B(V,V')$ we can consider $R(T)$ as a linear map T_R from R^{m+n} into $R^{m'+n'}$. Taking into account that $R(u)=R(v) \Rightarrow R(uT) = R(vT)$, there is a unique $T_R \in L(R^{m+n}, R^{m'+n'})$ making commutative the following diagram:

$$\begin{array}{ccc} V & \xrightarrow{\quad T \quad} & V' \\ {\scriptstyle R}\big\downarrow & & \big\downarrow{\scriptstyle R} \\ R^{m+n} & \xrightarrow{\quad T_R \quad} & R^{m'+n'} \end{array} \tag{4}$$

This defines the body map $R(T) = T_R$. The body T_R of a B-linear map represented by the matrix $A = (a^i_j)$ with entries in B, is represented by the matrix $R(A) = (r(a^i_j))$.

Theorem 3.4 Let $T \in L_B(V)$ and A be the matrix representation of T. Then i) if $R(A) = 0$, and $0 < \theta < 1$, there exists $\lambda > 0$ such that $\|A^n\| \leqslant \alpha\, \theta^n$ $\forall\, n \geqslant 0$. ii) T is a B-isomorphism iff T_R is a linear isomorphism. In addition, if $S(A) = A-R(A)$

$$A^{-1} = (R(A))^{-1} \sum_{n=0}^{\infty} (-1)^n \left(S(A)/_{R(A)^{-1}}\right)^n \tag{5}$$

Proposition 3.5 (dimension theorem) Given a B-isomorphism $T \in L_B(V,V')^0$ $(T \in L_B(V,V')^1)$ we have $m' = m$ and $n' = n$ $(m' = n$ and $n' = m)$.

Multi B-linear maps and B tensor and exterior products can be defined naturally in B-modules.

For details see ref. (1) and (5).

3. B. B-Spaces

Now we define B-spaces containing the superspaces as a subclass. In fact we are going to define a category whose objects are B-spaces and whose morphisms are L_B maps.

Definition 3.6 Given a m+n sequence $\pi = (p_1, p_2, \dots, p_{m+n})$ of β-projections of B. Let $V(\pi)$ be the Banach subspace of V, $V(\pi) = \{\sum_{i=1}^{m+n} a^i e^i \in V \mid a_i \in \operatorname{Im} p_i\}$. A B-space dimension is a couple $(\pi,\sigma) = ((p_i), (q_i))$ of m+n sequences of β-projections such that $q_i \cdot p_i = q_i$ (i.e. $\operatorname{Im} q_i \subset \operatorname{Im} p_i$). A B-space of dimension (π,σ) is the quotient Banach space $V(\pi,\sigma) = V(\pi)/_{V(\sigma)}$.

Definition 3.7 Let $V(\pi,\sigma)$ be a B-space, F_c a real vector space, and t a linear map from R^{m+n} onto F_c, then the linear map $t \cdot R/_{V(\pi)}$ from $V(\pi)$ into E factorizes to a linear map c from $V(\pi,\sigma)$ into F_c. We say that c is a k-body map of $V(\pi,\sigma)$ and that F_c is a body of $V(\pi,\sigma)$ if c is onto and k = dim F_c.

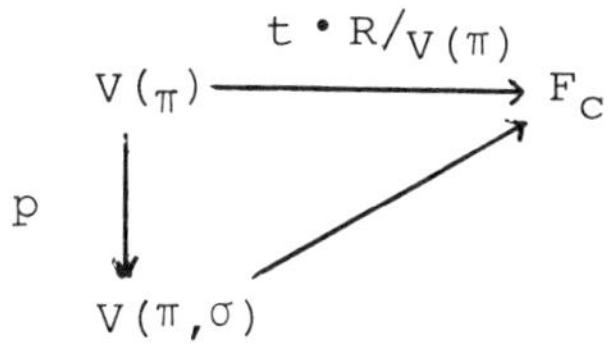

(6)

We now proceed to define B-linear operators between B-spaces.

Definition 3.8 Given $F = V(\pi,\sigma)$ and $F' = V'(\pi',\sigma')$ B-spaces and $T \in L(F,F')$ we say that T is L_B iff there exists $T' \in L(V(\pi), V(\pi))$ and $\overline{T} \in L_B(V,V')$ such that $pT = T'p'$ and $i\overline{T} = T'i'$ where i, i' and p, p' are respectively the inclusion and projections of the diagram below.

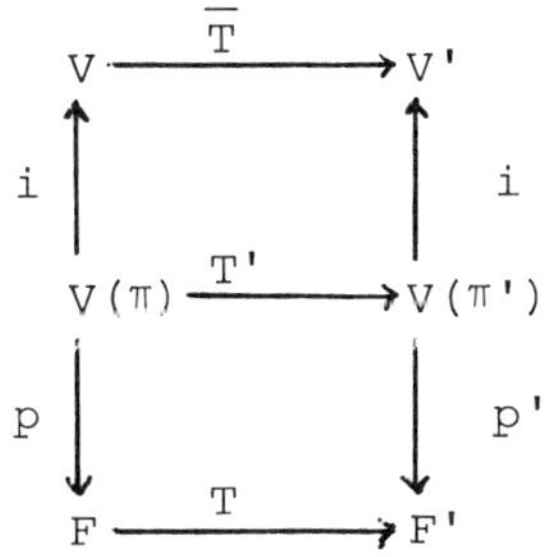

(7)

We call to $\overline{T}$ a factorizable extension of T.

Introducing

$$L_B(V,V';V(\pi),V'(\pi')) = \{T \quad L_B(V,V') \quad V(\pi)\; T \subset V'(\pi')\} \qquad (8)$$

We can endow $L_B(F,F')$ with a B-space structure through the identification

$$L_B(F,F') \approx \hat{L}_B(F,F') = L_B(V,V';F,F')/_{L_B(V,V';V(\pi),V(\sigma'))} \qquad (9)$$

where $L_B(V,V';F,F') = L_B(V,V';V(\pi),V(\pi')) \cap L_B(V,V';V(\sigma),V(\sigma'))$. (10)

Theorem 3.9 Let $F = V(\pi,\sigma)$, $F' = V'(\pi',\sigma')$ be B-spaces and $T \in L_B(F,F')$ a L_B-isomorphism (without defined parity). Then $\pi = \pi'$ and $\sigma = \sigma'$ up to a reordering of the basis of V'.

From now on we shall only consider even L_B-isomorphism.

As $L_B(F)$ is a B-space we can define on it many different bodies. However we will need the concept of body operator, as in 4.3, rather than that of the body itself, for the transition maps of fiber bundles. For a given body c of F we consider the subspace $c\text{-}L_B(F) = \{T \in L_B(F) \mid c(uT) = 0 \text{ if } c(u) = 0\}$. Then for each $T \in c\text{-}L_B(F)$ there exists a unique $T_c \in L(F_c)$ making commutative the following diagram:

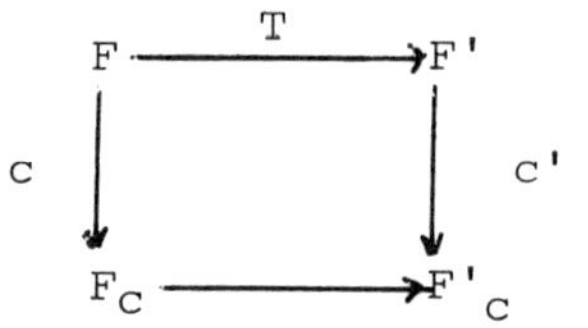

(11)

We call to T_c the c-body map of T and T is c-bodied. Of course there exist always a body on $L_B(F)$ such that its restriction to $c\text{-}L_B(F)$ produces T_c from T. We observe that if c is a maximal body then $c\text{-}L_B(F) = L_B(F)$. Similar considerations can be done for $L_B(F,F')$.

Given a B-space F modeled on V not all the basis of V make sense for F, but only those obtained from the canonical one $e_1, \ldots, e_{m+n}$ through an element of $GL_B(F)$. So $GL_B(F)$ can be considered as the set of reference frames on F. We remark that when F and/or F' are not free B-modules the space of B-linear maps from F into F' (in the usual sense) is bigger in general than the space $L_B(F,F')$.

To end up with the discussion about L_B maps between B-spaces, we define the dual B-space to a B-space F as $F^* = \hat{L}_B(F,B)$. It is easy to see that F is not reflexive in general ($F \subset F^{**}$). However F^* is always reflexive $F^{***} = F^*$.

3. C. Superspaces

Next we define superspaces as bodied B-spaces.

Definition 3.10 A superspace S is a B-space $S = V(\pi,\sigma)^0$ with $p_i = p_{F(K_i)}$, $K_i \subset N$ and $K_i \neq \emptyset$ $m+1 \leqslant i \leqslant m+n$. We consider in S the body r defined by $t: R^{m+n} \to R^m$, $t(x_1,\ldots x_{m+n}) = (x_1,\ldots x_m)$.

The superspace S is just the Banach subspace $\{\sum_{i=1}^{m+n} a^i e_i \in V / a^i \in B_{K_i}^{r_i}\}$ of V. So S is isomorphic to $B_K^0 \times\ldots\times B_{K_m}^0 \times\ldots\times B_{K_{m+1}}^1 \times\ldots\ldots \times B_{K_{m+n}}^1$. Then a superspace S has m even and n odd coordinates. Consequently we shall often use the notation $S^{m,n}$. Superspaces are a particular case of B-spaces and all the results of 3 B hold for superspace. However we outline some specific features concerning the dual space S* of superspace. S* is a B-module, (free iff S is free) and $S^{**} = \overline{S} = \{\sum_{i=1}^{m+n} a^i e_i \in V / a^i \beta_{K_i} = 0\}$ is also free. On the other hand $\overline{S}$ can be characterized as the maximal subset of V such that all extensions of maps in $L_B(S,V')$ coincide.

We observe that iff S is free, $L_B(S,S')$ is a superspace modeled on $L_B(V,V')$. Finally, we remember that a superspace $S^{m,n}$ has a canonical body map $r: S^{m,n} \to R^m$ which is maximal. Therefore $r',r\text{-}\hat{L}_B(S,S') = \hat{L}_B(S,S')$ and each $T \in L_B(S,S')$ has a unique $T_{r,r'} \in L(R^m)$ associated with it such that

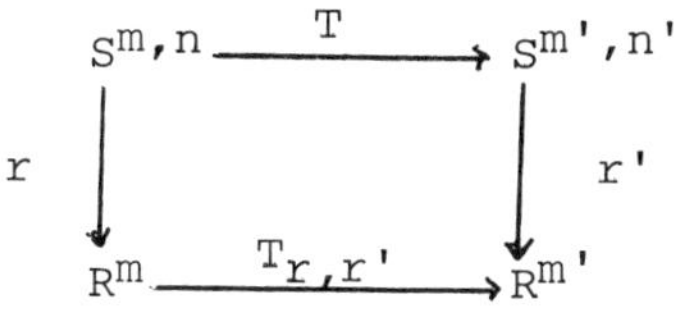

(12)

Concerning tensor and exterior algebra over a superspace, when S is free we have $S^* = V^*$ and $\overline{S} = \overline{V}$, so it reduces to the case of B-modules. For S not free we have the following situation. Since $\overline{S}$ is a (not-free) B-module, $S \otimes_B \overline{S}$ is determined by B-linearity and has the universal property.

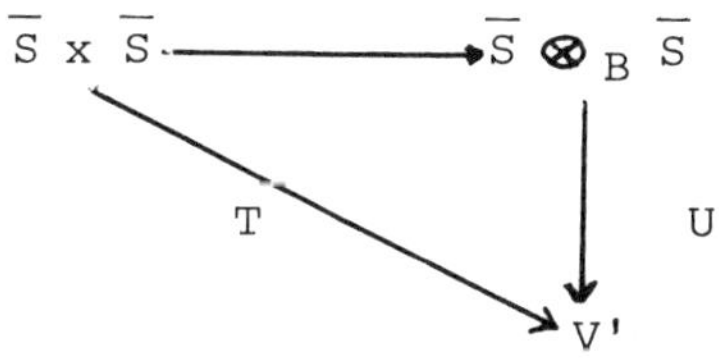

(13)

i.e.: for T B-bilinear there exists U B-linear making commutative the above diagram.However we are interested in L_B maps rather than in B-linear maps. As the last ones form in general a wider class than the former ones, we have that the space $\overline{S} \otimes_{L_B} \overline{S}$ satisfying the universal property for L_B maps is bigger than

$\overline{S} \otimes_B \overline{S}$. In fact $\overline{S} \otimes L_B \overline{S} \approx \hat{L}^2_B(S^*,S^*;B) \approx \overline{S} \otimes_B \overline{S}$. We remark that $\overline{S} \otimes L_B \overline{S}$ has also the universal property for L^2_B operators defined on $S \times S$. So $S \otimes_{L_B} S = \overline{S} \otimes_{L_B} \overline{S}$ and we shall denote it shortly by $S \otimes S$ or $\overline{S} \otimes \overline{S}$. Notice that first order tensors are then $\overline{S}$ and not S. These considerations can be straightfordwardly extended to higher tensor products of $\overline{S}$ and S^*.

4. ANALYSIS ON SUPERSPACES

We begin giving a notion of differentiation associated to L_B-linearity for functions defined in a domain U of a B-space F.

Definition 4.1 Given $f: U \to F'$, where F' is a B-space. i) f is G^1 if it is C^1 and there exist $\overline{D}f: U \to L_B(V,V')$ continuous such that $\overline{D}f$ is a factorizable extension of $Df(x)$ for each $x \in U$. ii) f is G^0 iff it is C^0. iii) f is G^k if it is G^1 and $\overline{Df}$ is G^{k-1}. iv) f is G^∞ if it is G^k for $k \geq 1$. v) f is G^ω iff it is G^∞ and analytic.

From definition follows immediately that if f is G^k differentiable then Df is G^{k-1} differentiable, and the composition of two G^k differentiable functions, f and g is G^k differentiable. A function $f: U \to V'$ defines the family $f^j: U \to B$, $1 \leq j \leq m'+n'$ by $f(\vec{x}) = \sum_{j=1}^{m'+n'} f^j(\vec{x})e'_j$, $\vec{x} \in U$. Then f is G^k-differentiable iff f^j, $1 \leq j \leq m'+n'$, is G^k-differentiable. The partial G-derivatives of a function appear in the following way.
For $f: U \to S'$(or V') $D^k f(\vec{x})$, $\vec{x} \in U$, is a L_B-multilinear map over S. A matrix representation of $D^k f(\vec{x})$ can be expressed as:

$$(h_1,\ldots,h_n)D^k f(\vec{x}) = \sum_{j=1}^{m'+n'} \Big(\sum_{1 \leq i_e \leq n+m} h_1^i \ldots h_k^{i_k} \, G_{i_k}\ldots G_{i_1} f^j(\vec{x})\Big)e'_j \tag{14}$$

that defines the partial G-derivatives $G_{i_k}\ldots G_{i_1} f^j(\vec{x})$. However, for S a non-free superspace, partial derivatives cannot be completely determined in that way. On the other hand if f is G^l the existence of partial derivatives $G_{i_k}\ldots G_{i_1} f$ that are G^{l-k} functions is guaranteed by the iterative form of definition 4.1.

Next we consider the set of G^k-functions from U into F'. In the case F' = B we denote this set by $G^k(U)$. $G^k(U)$ is a Z_2-graded commutative algebra and a Z_2 graded B-module. $G^k(U,F')$ is a Z_2 graded commutative $G^k(U)$-module. Leibnitz rule holds for derivations, (but only in a weak sense when S is not free)

$$\text{i.e.: } h_i G_i(fg) = h_i((G_i f)g + (-1)^{rr_i} f(G_i g)) \tag{15}$$

where r and r_i are the degrees of f and x^i respectively.

The main theorems of analysis in R^n are still true with the pertinent modifications. As an example we establish the inverse function theorem.

Theorem 4.2 Let $f\colon U \to F'$ be G^k, $k \geq 1$ and $\vec{x}_0 \in U$ such that $Df(\vec{x}_0)$ is an L_B-isomorphism. Then, there exists a neighbourhood of $\vec{x}_0$, $U_0 \subset U$ such that $f(U_0)$ is an open set of S' and $f\colon U_0 \to f(U_0)$ is a G^k-diffeomorphism (i.e.: one to one and f^{-1} is G^k).

We discuss now the extension of G^∞ functions and their natural domain of definition. We shall see that C^∞-functions behave along the soul as analytic functions. The extension theorem was first stated by Rogers (4), without fixing its domain of validity. Later, Jadczyk and Pilch (6) found the domain in which the theorem holds for homogeneous superspaces.

Let S be a superspace and $U \subset S$ a subset. The r-saturation of U is $\tilde{U} = r^{-1}(r(U))$. U is said G-connected (G-convex) if for all $\vec{x} \in U$, $\{\tilde{\vec{x}}\} \cap U$ is connected (convex). U open implies $\tilde{U}$ open.

Theorem 4.3 Let S be a superspace, $U \subset S$ a G-connected open set and $f \in G^\infty(U,F')$, where F' is a B-space with body c. Then i) If $\vec{x}, \vec{y} \in U$ and $r(\vec{x}) = r(\vec{y})$, then $cf(\vec{x}) = cf(\vec{y})$. ii) There exists a unique extension $\tilde{f} \in G^\infty(\tilde{U},F')$. iii) $c(\tilde{f}(\tilde{U})) = c(f(U))$. iv) Moreover, if $f \in G^\omega(U,F')$, then $\tilde{f} \in G^\omega(\tilde{U},F')$.

Corollary 4.4 Let $U \subset S$ be a G-connected open set and $U_0 \subset U$ an open set. Let $f, g \in G^\infty(U,F')$ such that $f = g$ in U_0. Then $f = g$ in $\tilde{U}_0 \cap U$.

Corollary 4.5 Let $U \subset S$ and $U' \subset S'$ be G-connected open sets and $f \in G^\infty(U,S')$, $g \in G^\infty(U',F'')$ with $f(U) \subset U'$. Then $\tilde{f}(\tilde{U}) \subset \tilde{U}'$ and $(gf)^\sim = \tilde{g}\cdot\tilde{f}$.

Corollary 4.6 Let $U \subset S$ be a G-connected open set and $f \in G^\infty(U,S')$ be a G^∞-diffeomorphism onto $f(U)$, G-connected open subset of S'. Then $\tilde{f} \in G^\infty(\tilde{U},S')$ is a G^∞ diffeomorphism onto $f(\tilde{U})$.

The need for the hypothesis $f(U)$ G-connected in corollary 4.6 will become clear in proposition 4.8. Prior to proposition 4.8 we introduce a map f_c associated to f by proposition 4.7.

Proposition 4.7 Let S, S' be superspaces and $U \subset S$, a G-connected open subset. Let $f \in G^\infty(U,F')$. Then there exist a unique function $f_c \in C^\infty(r(U);F'_c)$ such that $f_c \cdot r = c \cdot f$. This can be expressed by the commutative diagram.

$$\begin{array}{ccc} U & \xrightarrow{\quad f \quad} & F' \\ {\scriptstyle r}\downarrow & & \downarrow \\ r(U) & \xrightarrow{\quad f_c \quad} & F'_c \end{array} \qquad (16)$$

Proposition 4.8 Let U be a G-connected open subset of S and f a G^∞-diffeomorphism from U onto f(U) open in S'. i) If $V \subset r(U)$ is open and $f_r/_V$ is injective, then $f(r^{-1}(V) \cap U)$ is G-connected. ii) If $Z \subset f(U)$ is open G-connected, then $f_r/_{r \cdot f^{-1}(Z)}$ is injective. iii) For each $\vec{x} \in U$, $\exists\, \alpha > 0$ such that $f(B_\alpha(\vec{x})) \cap U$ is G-connected.

We conclude analysis in superspace giving the expansion of G^∞ functions in terms of C^∞ functions. Every G^∞ function can be expanded as a polinomyal in the odd coordinates where coefficients are functions of the even coordinates. By imposing some additional conditions these coefficients can be determined by functions of real variables. Moreover, as a consequence, partial G-derivatives will be determined. The following discussion is based mainly in Taylor expansion and the fact that the increments along the soul are almost nilpotent (nilpotents are dense in the soul).

We consider the case of $U = \tilde{U}$ a r-saturated open set of a superspace $S^{m,n}$ and the Z_2-graded algebra $G^\infty(U)$. Taking a point $p \equiv (x^i, \theta^j) \in \tilde{U}$, $1 \leqslant i \leqslant m$ and $m+1 \leqslant j \leqslant m+n$, we can expand a function $f \in G^\infty(U)$ around $(x^i, \vec{0})$ which does belong to $\tilde{U}$ because it is r-saturated. Then we obtain:

$$f(\vec{x}, \vec{\theta}) = \sum_{M \in F_n} \theta_M\, f_M(x) \qquad (17)$$

where $\theta_M = \theta_{i_1} \cdots \theta_{i_k}$, $\{i_1, \ldots, i_k\}$ is an ordered subset of $\{m+1, \ldots, m+n\}$ and F_n is the set of parts of $\{m+1, \ldots, m+n\}$. We are then lead to the study of the coefficients $f_M(x)$ defined in a superspace of type $S^{m,0}$. So we assume n = 0 in the following. Since U is r-saturated, $r(U) \subset U$ and a restriction map "Λ" can be defined from $G^\infty(U)$ into $C^\infty(r(U);B)$. Then the map $\Lambda: G^\infty(U) \to C^\infty(U)$ is linear injective and commute with partial derivatives i.e.: $G_i f = \partial\hat{f}/\partial x^i$. The map "$\Lambda$" is not in general surjective (unless card K_i is finite $\forall$ i) and we call $\hat{G}^\infty(U)$ to the image of $G^\infty(U)$ by "Λ" $G^\infty(U) = \mathrm{Im}$"$\Lambda$" $\subset C^\infty(U,B)$.

Now since "Λ" is injective we have the inverse linear operator Z from $\hat{G}^\infty(U)$ onto $G^\infty(U)$. Then Z commutes with derivatives

$$G_i(Z(g)) = Z\left(\frac{\partial g}{\partial x^i}\right) \qquad (18)$$

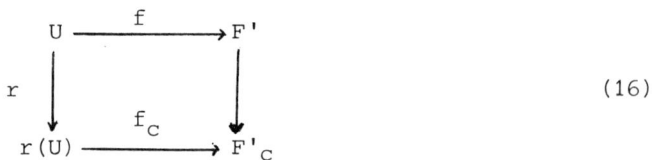

A characterization of $\hat{G}^\infty(U)$ is not easy to give in general, but as we show in the next theorem $\hat{G}^\infty(U)$ is wide enough to guarantee the existence of non-analytic C^∞ functions which in turn are needed to construct partitions of unity. Prior to state the next theorem we define $C^\infty_*(r(U);B)$ as the subset of functions $f \in C^\infty(r(U);B)$ such that their radii of convergence for their Taylor series are non-vanishing at each point $x \in r(U)$.

Theorem 4.9 a) For $g \in C^\infty_*(r(U);B)$, $g \in \hat{G}^\infty(U)$ and

$$Z(g)\ (i(x) + h) = \sum_{\alpha=0}^{\infty} \frac{1}{\alpha!} h^\alpha \frac{\partial^\alpha g}{\partial x^\alpha}(x) \tag{19}$$

for $x \in r(U)$, $r(h) = 0$ and $i(x) + h \in U$.
b) Moreover if $\cup K_i$ is finite, then $\hat{G}^\infty(U) = C^\infty(r(U);B)$ and (19) holds.

Next we construct "bell" functions on superspaces lifting by Z a bell function on R^m.

Corollary 4.10 Let S be a superspace and $U \subset S$ a G-connected open subset. Let $U_0 \subset U_1 \subset U$ be open subsets such that $r(U_0) \subset r(U_1)$ and is compact. Then, there exists $g \in G^\infty(U)$ such that $g = 1$ on $\tilde{U}_0 \cap U$ and $g = 0$ on $U - \tilde{U}_1$.

Concerning the properties of Z we have that Z preserves products and composition of C^∞ functions whenever it makes sense.

The following theorem establish precisely the expansion of G^∞ functions in terms of C^∞ functions.

Theorem 4.11 i) $f \in G^\infty(U)$, iff there exist a family $f_M \in \hat{G}^\infty(U)$; $M \in F_n$, such that

$$f(\vec{x},\vec{\theta}) = \sum_{M \in F_n} \theta_M Z(f_M)\ (x) \tag{20}$$

The expansion (20) (superfield expansion) allows a unique choice of partial G-derivatives for the case of non-free superspaces by the following recipe:

$$G_i f(\vec{x}) = \sum_{M \in F_n} \theta_M Z\left(\frac{\partial}{\partial x^i} f_M\right) (\vec{x}),\quad 1 \leqslant i \leqslant m \tag{21-a}$$

$$G_i f(\vec{x}) = \sum_{M \in F_n} \theta_M * \{i\} Z(f_M)\ (\vec{x}),\quad m+1 \leqslant i \leqslant m+n \tag{21-b}$$

where $\theta_M * \{i\} = 0$ if $i \notin M$ and $\theta_M * \{i\} = \pm\ \theta_{M-\{i\}}$ for $i \in M$. With this recipe $G_i f(\vec{x})$ are G^∞ if $f(\vec{x})$ is G^∞ and graded

Schwartz's lemma holds.

$$G_iG_jf(\vec{x}) = (-1)^{r_ir_j}G_jG_i\ f(\vec{x}) \tag{22}$$

Although the family f_L is not fixed and there are in general several superfield expansions for a given f when the superspace is not free, there is however a rule to assign to each G^∞ function a unique expansion, called the canonical expansion (5).

5. SUPERMANIFOLDS

We start with a concept of supermanifold which is a direct generalization of that of ref (2). However the existence of a well-behaved body for the supermanifolds impose restrictions on them. As a result of the analysis of these restrictions we arrive at the concept of ρ-supermanifolds.

Definition 5.1 Let S be a superspace and E a Banach C^∞ manifold (10) modeled on S. a) A G^∞ atlas on E is a C^∞ atlas $\{(U_\alpha,\psi_\alpha)\}_{\alpha\in\Lambda}$ on E, such that for each pair $\alpha,\beta \in \Lambda$, $\psi_{\alpha\beta} = \psi_\beta \cdot \psi_\alpha^{-1}$ is G^∞ from $\psi_\alpha(U_\alpha \cap U_\beta)$ onto $\psi_\beta(U_\alpha \cap U_\beta)$. b) A G^∞ structure on E is a maximal G^∞-atlas on E. c) A G^∞ supermanifold E is a Banach C^∞ manifold endowed with a G^∞ structure. If S is free we call E a free G^∞ supermanifold.

We analyze now how it is possible to associate a body $\rho: E \to E_\rho$ for the supermanifold, where E_ρ is a real C^∞ manifold. Also one should require that the map ρ in coordinates coincide with the map $r: S^{m,n} \to R^n$ when E is modeled on S.

As a first step we could establish the following equivalence relation between pairs of points $x,y \in E$. $x \sim y$ iff there exist a chart (U,ψ) in E such that $x,y \in U$ and $r\,\psi(x) = r\psi(y)$. However this relation is not useful in Haussdorff manifolds, because all points become related in a single class. Then a possible way to establish a non-trivial relation could be the following. $x \sim y$ iff there exist a cG-connected chart (U,ψ) such that $x,y \in V$ and $r\,\psi(x) = r\psi(y)$. It is possible however to construct a counterexample to show that this relation is not transitive (5). Nevertheless this difficulty can be circumvented by constructing a new relation by means of finite chains: $x \approx y$, iff there exist $x_1,\dots x_n$ E such that $x \sim x_1$, $x_1 \sim x_2,\dots,x_n \sim y$. The relation $\approx$ so obtained is yet not fully satisfactory by two reasons. First, there is no guarantee that $E/\approx$ be a real C^∞ manifold. Second, even if $E/\approx$ has a C^∞ manifold structure, the definition of body obtained by $\approx$ is not operative because two related points could have different bodies in a particular chart. On the other hand although G^∞ supermanifolds with compact soul and well defined body are possible G^∞ functions defined on them depend only on the body. In view of the above considerations we give a definition of ρ-supermanifolds which removes the difficulties.

Definition 5.2 Let E be a C^∞ manifold modeled on a superspace $S^{m,n}$, E_ρ a C^∞ manifold modeled on $\rho(S) = R^m$ and ρ a C^∞ map from E onto E_ρ. a) A ρ-atlas on the triple (E, E_ρ, ρ) is a pair of a G^∞ atlas $\{(U_\alpha, \psi_\alpha)\}$ on E and a C^∞ atlas $(U_{\rho_\alpha}, \psi_{\rho_\alpha})\}$ on E_ρ such that i) $\psi_\alpha(U_\alpha)$ is cG-connected. ii) $U_\alpha = \rho^{-1}(U_{\rho_\alpha})$. iii) $\psi_{\rho_\alpha} \bullet \rho = r \bullet \psi_\alpha$ (where r is the body map on $S^{m,n}$). b) A G^∞ ρ-structure on the triple (E, E_ρ, ρ) is a maximal ρ-atlas. c) A ρ-supermanifold is a triple (E, E_ρ, ρ) with a G^∞ ρ-structure on the triple.

From definition it follows that E_ρ is homomorphic to the body $E/\tilde{\sim}$. We shall not distinguish anymore between E_ρ and $E/\tilde{\sim}$. G^∞ functions between ρ-supermanifolds are defined as usual. Given two supermanifolds M, N and two G^∞ functions between them $f: M \to N$, there exists a unique body function $f_\rho: M_\rho \to N_\rho$ such that $\rho_N \bullet f = f \bullet \rho_M$.

From the existence of "bell" functions on superspaces, G^∞ partitions of unity in G^∞ supermanifolds can be inferred, although they are infinite dimensional C^∞ Banach manifolds.

Lemma 5.3 Let E be a ρ-supermanifold and $U \subset E$ an open set. Let $K \subset U$ be such that $\rho(K)$ is compact. Then there exists a neighbourhood U_0 of K, $U_0 \subset U$ and $g \in C^\infty(E)$ such that $g = 1$ on U_0 and $g = 0$ on $E-\tilde{U}$, and $0 \leq r \bullet g_\rho \leq 1$.

Theorem 5.4 Let E_ρ be paracompact and let $\{U_\alpha\}$ be an open covering of E. Then, there exists a partition of unity subordinate to $\tilde{U}_\alpha$, i.e.: there exists a family $\{\psi_\alpha\}, \psi_\alpha \in G^\infty(E)$ such that i) Supp $(\psi_\alpha) \subset \tilde{U}_\alpha$ ii) $\{$supp $(\psi_\alpha)\}$ is a locally finite covering of E, iii) $(\psi_\alpha)_\rho\ (x) \geq 0\ \forall x \in E_\rho$, iv) $\sum_\alpha \psi_\alpha\ (y) = 1$, $y \in E$.

The proof is based in the existence of partitions of unity in E_ρ, lemma 5.3 and the fact that Z is a linear map.

We discuss now vector fields on ρ-supermanifolds. It will emerge from the discussion that G^∞ vector fields do not coincide with G^∞ derivations, and two tangent spaces must be considered. First, we shall endow the usual tangent space T(E) over a ρ-supermanifold E with a ρ-structure. Let $T(E_\rho)$ be the tangent space of E_ρ (as Banach manifold), and $\rho_*: T(E) \to T(E_\rho)$ the derivative of ρ. Then, for $\{(U_\alpha, \psi_\alpha)\}$ a ρ-atlas of E $\{T(U_\alpha), \psi_{\alpha *}$ is a ρ-atlas for the triple $(T(E), T(E_\rho), \rho_*)$. This defines a ρ-supermanifold structure which does not depend on the specific

ρ-atlas chosen for E.

As T(E) is a ρ-supermanifold, it makes sense to consider G^∞ vector fields on E. Moreover, if X: E $\rightarrow$ T(E) is a G^∞ vector field, it is immediate, that X can be considered as a derivation on $G^\infty(E)$. However G^∞ vector fields do not exhaust all derivations on $G^\infty(E)$.

Definition 5.5 i) A supervector field X on E is a derivation of $G^\infty(E)$; i.e.: an element $X \in L_B(G^\infty(E))$ such that

$$X(fg) = (Xf)g + (-1)^{rs}f(Xg) \tag{23}$$

with $X \in L_B(G^\infty(E))^r$, $f \in G^\infty(E)^s$, $g \in G^\infty(E)$.
ii) A G^∞ vector field is a G^∞-section of T(E).

By lemma 5.3, if $U \subset E$ is open and $f \in G^\infty(E)$ vanish on U, then Xf = 0 for every supervector field X. Then, the field X in a chart (U,ψ) can be expressed by $X = \sum_{i=1}^{m+n} a^i\, \partial/\partial x^i$, where $\partial/\partial x^i f = G_i(f \cdot \psi^{-1})$ and $a^i: U \rightarrow B$ is a G^∞ function.

Proposition 5.6 Let (U,ψ) be a ρ-chart of E, then i) X is a supervector field on U iff $X = \sum_{i=1}^{m+n} a^i\, \partial/\partial x^i$, where $a^i \in G^\infty(U)$ and $\beta_{K_i}\, a^i = 0$, when $m+1 \leqslant i \leqslant m+n$, and K_i is finite. ii) X is a G^∞ vector on U iff X is a supervector field and $X = \sum_{i=1}^{m+n} a^i\, \partial/\partial x^i$ with $\mathrm{Im}\, a^i \subset B_{K_i}^{r_i}$, $r_i = 0$ for $1 \leqslant i \leqslant m$ and $r_i = 1$ for $m+1 \leqslant i \leqslant m+n$.

We shall denote the set of supervector fields, and G^∞ vector fields by $\overline{\chi}(E)$ and $\chi(E)$ respectively. We remark that the only supervector fields with associated (local) uniparametric group of transformations are the G^∞-vector fields, because each vector tangent to a curve on E is necessarily in T(E).

We observe that the space of supervector fields $\overline{\chi}(E)$ is a $G^\infty(E)$-module. It is also a Lie superalgebra (over B) with the bracket $[X,Y] = XY-(-1)^{|X||Y|}YX$. However $\chi(E)$ is not a $G^\infty(E)$-module, and the bracket gives it a structure of Lie algebra over R.

We analyze now the structure of $\chi(E)$ and $\overline{\chi}(E)$ as spaces of sections in some bundles. We introduce L_B-bundles as a generalization of vector bundles.

Definition 5.7 A L_B-bundle with base a ρ-supermanifold E and fiber a B-space F is a vector bundle $M(E,F,\pi)$ (7), such that there exists a trivializing covering $\{U_i,\tau_i\}$ satisfying that for each $x \in U_i$, the transition functions $x \rightarrow \psi_{ij}(x) = \tau_{ix} \cdot \tau_{jx}^{-1}$ from $U_i \cap U_j$ into $c\text{-}L_B(F)^0$ are G^∞ $\forall\, i,j$. From now on we shall refer to L_B-bundles as supervector bundles.

Given an L_B-bundle M and a body F_C for F, we can associate to them a real vector bundle M_C called the body of M, with base E_ρ, fiber F_C and transition functions $(\psi_C)_{ij}$ determined by the diagram:

$$\begin{array}{ccc} U_i \cap U_j & \xrightarrow{\ \psi\ } & c\text{-}G_{L_B}(F)^0 \\ \Big\downarrow \rho & & \Big\downarrow \overline{c} \\ \rho(U_i) \cap \rho(U_j) & \xrightarrow{\ \psi_C\ } & GL(F_C) \end{array} \tag{24}$$

Now given a ρ-supermanifold E, modeled on a superspace S, two tangent L_B bundles with fibers S and $\overline{S}$ can be considered. We design them by T(E) and $\overline{T}(E)$. Given a ρ-atlas $\{U_i, \psi_i\}$ of E, the transition functions of the tangent bundles are given by ψ_{ij*}. Taking the body r for S and $\overline{S}$, defined above, the body bundle (associated to r) of the two tangent bundles T(E) and $\overline{T}(E)$ is the usual tangent bundle $T(E_\rho)$. Also a cotangent bundle $T^*(E)$ with fiber $S^* = L_B(S,B)$ is constructed with transition functions ψ^*_{ij}. The body of $T^*(E)$ associated to the body r of S^* is the usual cotangent bundle $T^*(E_\rho)$.

The spaces of G^∞ vector fields $\chi(E)$, supervector fields $\overline{X}(E)$, and 1-superforms $\Omega(E)$, are the spaces of G^∞-sections of T(E), $\overline{T}(E)$ and $T^*(E)$ respectively. In general it is not possible to construct an extension of the bundle T(E) to an L_B-bundle with fiber V. The reason is that the transition functions $\psi_{\alpha\beta*}(x)$ have many extensions from S to V (unless S is free) and the choice of one extension satisfying the cocycle condition is not possible. When S is free, $\overline{S} = V$ and $\psi_{\alpha\beta}(x)_*$ has a unique extension from S to V.

Now we construct tensor bundles over a ρ-supermanifold as L_B-bundles. Given a ρ-supermanifold E, the tensor bundle $T^1_k(E)$ is an L_B-bundle with base E, fiber $T^1_k(V)$ and transition functions induced by ψ_{ij*}. These transition functions obviously have their image in $r\text{-}L_B(T^1_k(V))$. Associated with the body r of $T^1_k(V)$ we have a body $T^1_k(E)_r$ of $T^1_k(E)$ which coincides with $T^1_k(E_\rho)$.

Supertensor fields of type (1,k) are G^∞-sections of $T^1_k(E)$. A supertensor field T^1_k induces an ordinary tensor field $(T^1_k)_r$ on E_ρ, called the body of T^1_k.

Differential superforms of degree p are the G^∞ sections of the L_B-bundle $\Lambda_p(E)$. We denote the space of differential superforms by $\Omega(E)$. The exterior product in $\Omega(E)$ is defined through the exterior product in the fibers). The exterior differential d is an even linear map d: $\Omega_p(E) \to \Omega_{p+1}(E)$ satisfying i) $d(v \wedge w) = dv \wedge w + (-1)^p v \wedge dw$, $v \in \Omega_p(E)$ ii) $d^2 = 0$, iii) $df = \sum_{i=1}^{m+n} dx^i\, \partial f/\partial x^i$. We observe that d is L_B linear and that the Grassmann degree does not enter in the sign $(-1)^p$ in i) because d is even.

Differential superforms are obviously graded-skewsymmetric L_B-multilinear maps on supervector fields. As an example we give the exterior differential of a superform and the exterior product of two superforms.

$$(X_1,X_2)d\omega_1 = X_1(X_2\omega_1) - (-1)^{(r_1+s_1)s_2} X_2(X_1\omega_1) - ([X_1,X_2])\omega_1 \tag{25}$$

$$(X_1,X_2)\ \omega_1 \wedge \omega_2 = (-1)^{r_2 s_1}(X_1\omega_1)(X_2\omega_2) + (-1)^{1+r_1(r_2+s_1)}(X_2\omega_1)(X_1\omega_2) \tag{26}$$

where $X_i \in \overline{X}(E)^{r_i}$ and $\omega_i \in \Omega_1(E)^{s_i}$.

For a G^∞ function between two ρ-supermanifolds $f: M \to N$, the usual properties $f^*(\omega_1 \wedge \omega_2) = f^*\omega_1 \wedge f^*\omega_2$ and $f^*d\omega = df^*\omega$ hold.

6. SUPERGROUPS AND PRINCIPAL SUPERFIBER BUNDLES

We describe in this section how supergroups and principal superfiber bundles are included within the framework of ρ-supermanifolds. We shall restrict ourselves to the case of free ρ-supermanifolds.

Definition 6.1 A Lie supergroup G is a ρ-supermanifold with a group structure given by a composition law such that for $a,b \in G$ the map $(a,b) \to ab^{-1}$ is G^∞.

As a consequence of definition 6.1, the body G_ρ of G is a Lie group. The composition law for G_ρ is induced from that of G by $\rho(a)\,\rho(b) = \rho(ab)$. So the body map ρ is a group homomorphism $\rho: G \to G_\rho$.

Left and right translations L_a and R_a are G^∞-diffeomorphisms. Then, the space of left invariant supervector fields $\overline{\mathfrak{g}}$ is a lie superalgebra over B, i.e.: the Lie bracket is graded and B-bilinear. On the other hand the space of left invariant G^∞ vector fields is a B-lie algebra, i.e.: the Lie bracket involves only commutators and is L_B-bilinear.

There exists an L_B-isomorphism from $\overline{\mathfrak{g}}$ onto $\overline{T}_e(G)$, that takes $\mathfrak{g}$ onto $T_e(G)$. Then the B-Lie algebra $T_e(G)$ and the

B-Lie superalgebra $\overline{T}_e(G)$ are B-spaces with body given by the real Lie algebra $\mathfrak{G}_\rho$ of G_ρ.

Another body can be defined for $\overline{\mathfrak{G}}$ in the following way. In a basis $\{X_1, \ldots, X_{m+n}\}$ of $\overline{\mathfrak{G}}$ the Lie superalgebra is determined by the structure constants $[X_i, X_j] = \pm c^k_{ij} X_k$. Take now $V = V^0 \oplus V^1$ to be a real graded vector space with $\dim V^0 = m$ and $\dim V^1 = n$. Choose a basis $\{Y_1, \ldots, Y_{m+n}\}$ with $\{Y_1, \ldots Y_m \in V^0$ and $Y_{m+1} \ldots Y_{m+n} \in V^1$. We define R: $\overline{\mathfrak{G}} \to V$ by $R(\sum_{i=1}^{m+n} a^i X^i) = \sum_{i=1}^{m+n} r(a^i) Y_i$. Then V is furnished with a real Lie superalgebra structure by the bracket $[Y_i, Y_j] = r(c^k_{ij}) Y_k$. This provides an ordinary real graded Lie algebra (8) associated to the group G which we shall call $\mathfrak{G}_R$. In addition we have that $\mathfrak{G}_\rho$ is isomorphic to $\mathfrak{G}^0_R$.

The elements of $\mathfrak{G}$ give rise to real monoparametric Lie subgroups of G while for elements $X \in \overline{\mathfrak{G}}$ such that $X \notin \mathfrak{G}$, an integral curve does not exist. Hence, the exponential maps $\mathfrak{G}$ into G, but it is not defined on the whole $\overline{\mathfrak{G}}$.

Definition 6.2 A matrix representation D of dimension (m,n) of a supergroup G is a G^∞ homomorphism D: $G \to GL_B(V)^0$, where V is a free B-module of dimension (m,n).

Given a matrix representation:

$$D(g) = \begin{pmatrix} D_{11} & D_{12} \\ D_{21} & D_{22} \end{pmatrix} (g) \tag{27}$$

of a supergroup G, $D_{11}(g)$ and $D_{22}(g)$ are m x m and n x n dimensional respectively and both have entries in B^0. On the other hand $D_{12}(g)$ and $D_{21}(g)$ are respectively m x n and n x m dimensional and both have entries in B^1. Since D is G^∞

$$\rho(g) = \rho(h) \Rightarrow D(g)_R = D(h)_R = \begin{pmatrix} R(D_{11}(g)) & 0 \\ 0 & R(D_{22}(g)) \end{pmatrix} \tag{28}$$

So a representation D of G induces a reducible real (m+n)-dimensional representation of G_ρ, D_R: $G_\rho \to GL(R^{m+n})$ given by: $D(\rho(g)) = D(g)_R$. Then D_R reduces to $D_R = D_\rho \oplus D'_\rho$, where $D_\rho : G_\rho \to GL(R^m)$, $D'_\rho : G_\rho \to GL(R^n)$ and $D_\rho(\rho(g)) = R(D_{11}(g))$, $D'_\rho(\rho(g)) = R(D_{22}(g))$. We remark that the matrix supergroups of Rittenberg and Scheunert (9) are examples of the G^∞ supergroups defined in 6.1, whose body is a direct product G x G .

Given a representation D of a supergroup we have, associated to it, straightforwardly, representations for the "Lie" algebras $\overline{\mathfrak{G}}$, $\mathfrak{G}$, $\mathfrak{G}_R$ and $\mathfrak{G}_\rho$ associated with G.

Definition 6.3 A principal superfiber bundle P is a principal fiber bundle P(M,G,π) such that P and M are ρ-supermanifolds, G is a Lie supergroup, and π , the transition functions, and the trivializations are G^∞ functions.

As a consequence of the definition, P_ρ is a principal fiber bundle P_ρ $(M_\rho, G_\rho, \pi_\rho)$. The L_B-bundles defined in section 5 with base E and fiber a B-space F, are associated to a principal superfiber bundle $P(E, GL_B(F))$.

Associated to an element $A \in \mathfrak{g}$, we have a fundamental G^∞-vector field A^* constructed by means of the monoparametric subgroup of A. To deal with connections and curvature forms, it is convenient to extend the fundamental fields to elements of $\bar{\mathfrak{g}}$. To do that, we look at $A_u^* = \sigma_{u\,*\,e}(A)$ where $\sigma_u: G \to P$ is given by $\sigma_u(g) = ug$. The last expression for A_u^* makes a sense also for elements of $\bar{\mathfrak{g}}$, and $A_u^* \in \bar{T}^*(P)$.

A superconnection form on a principal superfiber bundle P, is defined as an even element ω of $\Omega^1_{\bar{\mathfrak{g}}}(P)$ i.e.: a $\bar{\mathfrak{g}}$-valued even superform on P satisfying the conditions: $R_a^*\omega = ad(a^{-1})\omega$ $\forall a \in G$ and $A^*\omega = A$ $\forall A \in \bar{\mathfrak{g}}$.

The vertical part of $\bar{T}_u(P)$ is determined, as usual, as the tangent part to the fiber $\pi^{-1}(u)$ on u. $\bar{T}(\pi^{-1}(u)) = \bar{G}_u$ and $\sigma_{u\,*\,e}$ gives a B-isomorphism from $\bar{\mathfrak{g}}$ into $\bar{G}_u$. The horizontal part for ω is $h(\bar{T}_u(P)) = \mathrm{Ker}\,\omega(u)$, and $\bar{T}_u(P) = h\,\bar{T}_u(P) \oplus \bar{G}_u$. So h is a well defined G^∞ projection.

The covariant differential D is defined by applying the exterior differential to the horizontal part of the super-vector fields. So the curvature $\Omega(X,Y) = D\omega(X,Y) = d\omega(hX,hY)$ is a G^∞ $\bar{\mathfrak{g}}$-valued even two superform, The structure equation $\Omega = d\omega + [\omega,\omega]$ and the Bianchi identity $D\Omega = 0$, hold.

REFERENCES

(1) F.A. Berezin, D.A. Leites, Sov. Math. Dokl. 16, 1218, (1975).
D.A. Leites, Uspekhi Mat. Nauk. 35, 1, (1980) (Russian Math. Surveys 35, 1, (1980)

(2) B. Kostant, in Differential Geometrical Methods in Mathematical Physics, Bonn 1975 (Springer-Verlag, Berlin 1977)

(3) J. Dell, L. Smolin. Comm. Math. Phys. 66, 197, (1979).

(4) A. Rogers, Journ. Math. Phys. 21, 1351, (1980)

(5) J. Hoyos, M. Quirós, J. Ramírez-Mittelbrunn, F.J. de Urries Generalized Supermanifolds I, II, III (to appear in Journ.

Math. Phys.)

(6) A. Jadczyk, K. Pilch. Comm. Math. Phys. 78, 373, (1981)

(7) S. Kobayashi, K. Nomizu. Foundations of Differential Geometry. Vol I. Interscience. New York, 1963

(8) M. Scheunert. The Theory of Lie Superalgebras Lecture Notes in Mathematics Nº 716. Springer Verlag. Berlin, 1979

(9) I. Rittenberg, M. Scheunert. Journ. Math. Phys. 19, 709, (1978).

(10) S. Lang. Differential Manifolds. Addison Wesley. Reading Mass. (1972).

A LAGRANGIAN FOR SU(2/1) QUANTUM ASTHENODYNAMICS

Jean Thierry-Mieg* and Yuval Ne'eman+

*Groupe d'Astrophysique Relativiste CNRS
Observatoire de Meudon 92190, France
+Tel Aviv University, Tel Aviv, Israel
and University of Texas, Austin, Texas, USA

ABSTRACT

A formally unitary Lagrangian model for an internal supersymmetry is constructed by gauging the even solvable Lie algebra appearing as the even part of the product of the simple superalgebra with the Grassmann Manifold of forms over space-time. The construction explains such features as matter-field ghosts, the correlation of chirality with the superalgebra's representations' gradings, the Kronecker metric and the vector ghosts in the adjoint representation.

1. It was suggested (1,2) that all of the apparently abitrary selections of $[SU(2)xU(1)]_{W.E.}$ quantum numbers for the matter and Goldstone-Higgs fields in the Weinberg-Salam unified theory of weak and electromagnectic interactions ("asthenodynamics") could be explained by postulating the simple supergroup SU(2/1) as a constraining structure. Moreover, this approach can be further extended to include $SU(3)_{color}$ and the observed sequential pattern of quark/lepton "generations" (3). A model based (4) on SU(7/1) predicts eight generations, four of them with inverted chiralities. This happens to correspond (16 "flavours") to "critical QCD" and is said to explain uniquely the observed features of high energy scattering (5).

The manner in which a supergroup acts internally is not clear, as the graded vector spaces carrying its action should be bringing together states with identical behaviour under the Lorentz group, but opposite Quantum statistics. We have conjectured that the $\mathfrak{su}(2/1)$ superalgebra transitions relate particle fields to ghost fields, of the type introduced by Feynman, DeWitt, Faddeev and Popov to ensure the Unitarity of

S. Sternberg (ed.), Differential Geometric Methods in Mathematical Physics, 101–114.

Quantized Gauge Theories. $su(2/1)$ thus resembles the Becchi-Rouet-Stora (6) and the Curci-Ferrari (7) algebras, also relating particle and ghost fields, and providing algebraic shortcuts, simplifying the imposition of Unitarity throughout the renormalization procedure.

2. SU(2/1) involves several unusual features:

a) The grading of the carrier vector-spaces in the (spinorial) matter representations appears to relate to chirality. Indeed, we have proved elsewhere (8) that the (reducible)matrices representing $[SU(2)xU(1)]_{W.E.}x\ SU(3)_{color}$, when acting on supermultiplets constructed from juxtapositions of independent matter multiplets and graded according to chirality (thus disregarding Lorentz invariance in that auxiliary construction) have vanishing supertraces. It is thus not surprising that they should in that form be embeddable in Unitary Superunimodular groups such as SU(2/1) or SU(7/1), whose generator superalgebras are supertraceless by definition. However, if we require $su(2/1)$ or $su(7/1)$ to commute with the Lorentz group, we have to replace the particles in either one of the gradings by ghosts (to ensure a change of statistics under the action of the odd part of the superalgebra) with inverted chiralities. For example, the two $[SU(2)_I x\ U(1)_U]_{weak}$ multiplets of e-leptons $(e\nu^o_L, e^-_L)$ with $I=\frac{1}{2}$, $U=-1$ and (e^-_R) with $I=0$, $U=-2$ can be combined as a reducible $({}_e\nu^o{}_L, e^-{}_L)_1 \oplus (e^-{}_R)_o$, the external index defining a grading (as yet without the action of a superalgebra). The matrices of $su(2)_I$ act only on the odd-graded left-chiral doublet, and are traceless; the $u(1)_W$ matrix has equal traces -2 in both gradings. The difference between the traces of the two gradings thus vanishes for the entire $su(2)\ x\ u(1)$. However, when these matrices are embedded in $su(2/1)$, the superalgebra's (irreducible) 3-dimensional carrier space has to be $({}_e\nu^o_L,\ e^-_L\ /\ X_L(e^-_R))$, where $X_L(e^-_R)$ is a left-chiral ghost field with the quantum numbers of the right-chiral electron $e^-{}_R$. Whether such previously unencountered $X_L(e^-{}_R)$ ghosts were only artificial constructs or had a "physical" role (in the sense of playing a role in the dynamical quantized theory), and how they reflect on the counting of (true) physical states and unitarity - these were two of the new puzzles introduced by the use of a supergroup internally. The second part of this puzzle was resolved in a study (9) in which we constructed the relevant Curci-Ferrari algebra for the SU(2/1) gauge and matter multiplets and showed that the ghost-antighost doubling occurring in the Curci-Ferrari construction ensures that for each state in the matter multiplet one gets a total of $|1 - 1 - 1| = 1$ physical degrees of freedom (the ghost and antighost count in the negative). the model we display in the present work will answer the first part of the puzzle.

b) In the $\underset{\sim}{8}$ adjoint representation, a most satisfactory feature is the appearance, for spin J=0 of the four SU(2)xU(1) Faddeev-Popov ghosts α together with the Goldstone-Higgs field h. However, for J=1, the intermediate bosons (and electromagnetic field) A_μ are accompanied by yet another previously unencountered vector-ghost field β_μ. The conventional Weinberg-Salam lagrangian (with spontaneous symmetry breakdown) was shown by 't Hooft and others to produce a unitary theory without such β_μ ghosts. Is there a version of the theory requiring it?

c) The supergroup matrices predict a Weinberg angle $\theta_W = 30^o$, provided they are normalized by their traces, and using a Kronecker δ metric (as in SU(3)) rather than the SU(2/1) Killing metric, which is not positive definite.

3. The exponentiation of a superalgebra $g=g^+ + g^-$ (the even and odd parts) requires (10) the parameters to belong to the even and odd parts of a Grassmann algebra $\Omega = \Omega^+ + \Omega^-$:

$$G(g,\Omega) = \exp\ (\Omega^+ \otimes g^+ + \Omega^- \otimes g^-) \qquad [3.1]$$

The supergroup G thus depends on the dimensionality of the generating basis Ω^1 of Ω; in fact it can be said that a supergroup is a Lie superalgebra valued Grassmann manifold. We see in (3.1) that it can also be generated by the exponentiation of a (reducible) Lie algebra $\tilde{g}$

$$\tilde{g} = \Omega^+ \otimes g^+ \otimes \Omega^- \otimes g^- \qquad [3.2]$$

whose dimensionality $[\tilde{g}]$ depends on that of Ω^1,

$$[\tilde{g}] = \tfrac{1}{2}\ [g] \times 2^{[\Omega^1]} \qquad [3.3]$$

In our dynamical realization (11) of G=SU(2/1) we work with a specific Ω, the Grassmann algebra of forms over space-time Grassmann's original algebra at the time) $\Omega(M_{3.1})$. The corresponding gauge theory is therefore highly soldered to the base space in the gauge bundle. Most of the attractive properties of differential geometry are maintained and our construction provides a non-trivial example of a Cartan Integrable System (12). The generators of $\tilde{g}$ will be denoted as $\lambda_a = \mu_a$, $\lambda_i{}^\mu = \mu_i\ dx^\mu$, $\lambda_a{}^{\mu\nu} = \mu_a\ dx^\mu \wedge dx^\nu, \ldots$ $(\mu_a \leftarrow g^+ , \mu_i \leftarrow g^-)$

The Lie bracket is defined as the exterior product for the forms, times the Lie superbracket $[\mu_M, \mu_N\}$. For instance

$$[\lambda_i^{\ \mu}, \lambda_j^{\ \nu}\} = dx^\mu \wedge dx^\nu \{\mu_i . \mu_j\} = f_i^{\ \mu}{}_j^{\ \nu}{}_{\rho\ \sigma}^{\ a} \lambda_a^{\ \rho\sigma} \qquad [3.4]$$

The Jacobi identity is automatically satisfied. Throughout this paper we use square brackets to denote the even subgroup SU(2)xU(1) commutation relations, and curly brackets to denote those relations originating specifically in the superalgebra SU(2/1). The brackets really denote commutators and anticommutators of number matrices, once all exterior products have been evaluated.

Such a generalized gauge theory involves a generalized system of connections, skew-symmetric contravariant, Bose tensor gauge fields $A_\mu{}^a$, $B_{\mu\nu}{}^i$, $C_{\mu\nu\rho}{}^a$, $E_{\mu\nu\rho\sigma}{}^i$ of alternating supergroup gradings, saturating the dimensionality of space-time forms. Under an infinitesimal transformation with parameter $\tilde{\varepsilon}$ $(\varepsilon^a, \varepsilon_\mu{}^i, \varepsilon_{\mu\nu\rho}{}^i, \varepsilon_{\mu\nu\rho\sigma}{}^a)$, the gauge fields vary according to (D denotes the λ_a covariant differential with gauge field A^a_μ; d is the exterior differential).

$$\begin{aligned}
\delta A &= -D\,\varepsilon^{(o)} := -\,d\,\varepsilon^{(o)} - [A, \varepsilon^{(o)}] &&=: \tilde{D}\varepsilon^{(o)} \\
\delta B &= -D\,\varepsilon^{(1)} - [B, \varepsilon^{(o)}] &&=: \tilde{D}\varepsilon^{(1)} \\
\delta C &= -D\,\varepsilon^{(2)} - \{B, \varepsilon^{(1)}\} - [C, \varepsilon^{(o)}] &&=: \tilde{D}\varepsilon^{(2)} \\
\delta E &= -D\,\varepsilon^{(3)} - [B, \varepsilon^{(2)}] - [C, \varepsilon^{(1)}] - [E, \varepsilon^{(o)}] &&=: \tilde{D}\varepsilon^{(3)}
\end{aligned} \qquad [3.5]$$

where $A = A_\mu{}^a \mu_a dx^\mu$, $B = \frac{1}{2} B_{\mu\nu}{}^i \mu_i dx^\mu \wedge dx^\nu$, $C = {}^1/_6\, C_{\mu\nu\rho}{}^a dx^\mu \wedge dx^\nu \wedge dx^\rho$, $E = {}^1/_{24}\, E_{\mu\nu\rho\sigma}{}^i \mu_i dx^\mu \wedge dx^\nu \wedge dx^\rho \wedge dx^\sigma$, and exterior products are implied.

These equations define the action of the generalized covariant derivative $\tilde{D}$. The generalized curvature $\tilde{F}$ is similarly defined,

$\tilde{F}$ (F^a, G^i, H^a)

$$\begin{aligned}
F^a &= dA + \tfrac{1}{2}[A, a]^a \\
G^i &= (DB)^i \\
H^a &= (DC)^a + \tfrac{1}{2}\{B, B\}^a
\end{aligned} \qquad [3.6]$$

these curvatures transform covariantly,

$$\delta\tilde{F} = [\tilde{\varepsilon},\tilde{F}] : \delta F = -[F,\varepsilon^{(o)}]$$

$$\delta G = -[G,\varepsilon^{(o)}] - [F,\varepsilon^{(1)}]$$

$$\delta H = -[H,\varepsilon^{(o)}] - \{G,\varepsilon^{(1)}\} - [F,\varepsilon^{(2)}] \qquad [3.7]$$

and satisfy the Bianchi identity

$$\tilde{D}\tilde{F} = o : DF = o$$

$$DG + [B,F] = o \qquad [3.8]$$

4. The matter multiplets are similarly treated. An irreducible representation $R = R^+ + R^-$ of the superalgebra g will give rise to a representation $\tilde{R}$ of $\tilde{g}$. Denote by $\tilde{\phi}$ a system of 0,1,2... forms taking their values alternatively in R^+ and R^-, $\tilde{\phi}$ $(\phi^A . \psi_\mu{}^I dx^\mu, \Xi_{\mu\nu}{}^A dx^\mu \wedge dx^\nu, ..)$. The representation is defined by the transformation rules,

$$\delta\ \phi = [\varepsilon^{(o)}, \phi]$$

$$\delta\ \psi = [\varepsilon^{(o)}, \phi] + \{\varepsilon^{(1)}, \phi\} \qquad [4.1]$$

$$\delta\ \Xi = [\varepsilon^{(o)}, \Xi] + \{\varepsilon^{(1)}, \psi\} + [\varepsilon^{(2)}, \phi]$$

with the Jacobi identity automatically satisfied. In the SU(2/1) system, the connections [3.5] are the generalized gauge fields and the curvature [3.6] their field-strengths. For the leptons (and quarks) we have for $\tilde{g}$ a doublet left-spinor $(\phi)_L$ together with a singlet (two singlets) left-vector-spinor $(\phi_\mu)_L$ etc..., all with Fermi statistics (but increase of spin along the representation, since the parameters $\tilde{\varepsilon}$ are not Lorentz scalars).

5. We now exhibit a free field Lagrangian such that it has precisely the same physical degrees of freedom as the Weinberg-Salam model. Given a skew p tensor $\psi_{\mu\nu...}$ and its exterior derivative (or generalized curl) this is

$$L_p = -\frac{1}{2p!}\ (\delta_{[\mu}\ \phi_{\nu\rho..]})^2 \qquad [5.1]$$

For a scalar ϕ, this is the Klein-Gordon equation; for ϕ_μ, this is Maxwell's Lagrangian; for $\phi_{\mu\nu}$ this is Kalb Ramond (13,14). In

N dimensions (N-1 space-type, 1 time) and for the massless Lagrangian [5.1], there is a duality equivalence (15) between p and N-p-2 forms (this is Hodge duality in the transverse dimensions).

It can be shown that the number of physical degrees of freedom n for the gauge fields [3.5] is precisely given by adding up the number $\binom{N}{k}$ of components of an antisymmetric k-indices tensor in N dimensions, together with the number of dy-contracted components of its complexified (7,9,16,17) geometrical vertical complements (in the direction of the fiber y^M, in the bundle manifold, ghosts counting negatively). For the forms [3.5], denoting the fiber-complexified forms by a caret.

$$\hat{A}^a = A^a_\mu\, dx^\mu + A^a_M\, dy^M + A^d_{\bar{N}}\, dy^{\bar{N}} = A^A_\mu\, dx^\mu + \alpha^a + \bar{\alpha}^a$$

$$\hat{B}^i = \tfrac{1}{2} B^{\ i}_{\mu\nu}\, dx^\mu \wedge dx^\nu + B^i_{\mu M}\, dx^\mu \wedge dy^M + B^i_{\mu\bar{N}}\, dx^\mu \wedge dy^{\bar{N}}$$

$$+ \tfrac{1}{2} B^i_{MN}\, dy^M \wedge dy^N + B^i_{M\bar{N}}\, dy^M \wedge dy^{\bar{N}} + \tfrac{1}{2} B^i_{\bar{M}\bar{N}}\, dy^{\bar{M}} \wedge dy^{\bar{N}} \qquad [5.2]$$

$$= \tfrac{1}{2} B^{\ i}_{\mu\nu}\, dx^\mu \wedge dx^\nu + \beta^i_\mu\, dx^\mu + \bar{\beta}^i_\mu\, dx^\mu + b^i + h^i + \bar{b}^i$$

$$\hat{C}^a = 1/6\ C^a_{\mu\nu\rho}\, dx^\mu \wedge dx^\nu \wedge dx^\rho + \tfrac{1}{2} C^a_{\mu\nu M}\, dx^\mu \wedge dx^\nu \wedge dy^M +$$

$$\tfrac{1}{2} C^a_{\mu\nu\bar{N}}\, dx^\mu \wedge dx^\nu \wedge dy^{\bar{N}} + \tfrac{1}{2} C^a_{\mu MN}\, dx^\mu \wedge dy^M \wedge dy^N +$$

$$C^a_{\mu M\bar{N}}\, dx^\mu \wedge dy^M \wedge dy^{\bar{N}} + \tfrac{1}{2} C^a_{\mu\bar{M}\bar{N}}\, dx^\mu \wedge dy^{\bar{M}} \wedge dy^{\bar{N}} +$$

$$1/6\ C^a_{MNP}\, dy^M \wedge dy^N \wedge dy^P + \tfrac{1}{2} C^a_{MN\bar{P}}\, dy^M \wedge dy^N \wedge dy^{\bar{P}} +$$

$$\tfrac{1}{2} C^a_{M\bar{N}\bar{P}}\, dy^M \wedge dy^{\bar{N}} \wedge dy^{\bar{P}} + 1/6\ C^a_{\bar{M}\bar{N}\bar{P}}\, dy^{\bar{M}} \wedge dy^{\bar{N}} \wedge dy^{\bar{P}}$$

$$= {}^1/6\ C^a_{\mu\nu\rho}\, dx^\mu \wedge dx^\nu \wedge dx^\rho + \tfrac{1}{2} (\Gamma^a_{\mu\nu} + \bar{\Gamma}^a_{\mu\nu})\, dx^\mu \wedge dx^\nu +$$

$$(C^{a+}_\mu + C^a_\mu + C^{a-}_\mu)\, dx^\mu + \gamma^{a---} + \gamma^{a+} + \gamma^{a+++}$$

etc.

Latin letters denote Bose fields; Greek, Fermi ghost fields. We have (per internal index)

$$n(A) = \binom{4}{1} - (2 \times 1) = 2$$

$$n(B) = \binom{4}{2} - (2 \times 4) + 3 = 1$$

$$n(C) = \binom{4}{3} - (2 \times 6) + (3 \times 4) - 4 = o \qquad [5.3]$$

$$n(E) = \binom{4}{4} - (2 \times 4) + (3 \times 6) - (4 \times 4) + 5 = o$$

We note that $\hat{B}^i$ contains a scalar real h^i multiplet, required in the SU(2/1) irreps $\mathfrak{g}^1(\alpha^a, h^i)$ or $\bar{\mathfrak{g}}^1(\bar{\alpha}^a, h^i)$, in the Curci-Ferrari type symmetric-complexified algebra of ghosts (7). The higher forms C and E do not contribute to the physical spectrum, nor would the total contribution of the system of non vanishing ghosts of a higher tensor.

The origin of this method of counting is derived from a geometrical interpretation of the ghost fields which we introduced several years ago (18-23). We showed that ghosts may be identified with the contracted vertical piece of the connection on a Principal Fiber Bundle, once one has picked a section, to define a local field. The BRS equations are then just the Maurer-Cartan equations describing the geometry of the bundle. They state that the Curvature 2-form is a horizontal object, i.e. has no components multiplying $dy^m \wedge dy^n$ or $dx^\mu \wedge dy^m$, whatever the choice of section. The role of the antighost was still unclear, until M. Quiros and collaborators (24) extended the construction to a doubled Principal Bundle $P_2(M, G \times G)$ which is equivalent to the complex construction in dy and $d\bar{y}$ that we use here, They showed that the Cartan-Maurer equations on that doubled bundle reproduce the algebra of Curci and Ferrari (7).

Alternative interpretations have followed. Bonora and Tonin (25) have postulated a superspace in which Minkowski $M_{3,1}$ is extended by two anticommuting variables $\theta, \bar{\theta}$. This amounts to projecting all our m or $\bar{m}$ indexed contractions with odd numbers of dy^m or $d\bar{y}^{\bar{m}}$ onto a basis $\theta, \bar{\theta}$. Delbourgo and Jarvis (26) have extended the algebraic structures on that superspace to the full OSp(4/2).

Some authors have found fault with the geometric interpretation (27), claiming that the product of 1-forms on the bundle is nilpotent, a fact which would impair the application of the formalism to path-integrals, in which there should appear an infinite product $\prod_x \alpha(x)$ of ghosts. The authors have perhaps indeed thereby detected the lack of a mathematically detailed precise definition of that product in reference (18-22). It is, however, rather trivially clear that the cotangent space to the bundle is infinite. Thus, the exterior product needed to realize

$\prod_x$ has to be taken over that infinite space. The definition supplied in equation [15] of reference (27) is indeed that of the vertical part of the connection, once a section s has been picked (whereas equation [12] is generally wrong, requiring either trivialization of the bundle or a special choice of co-ordinates in order to realize non-dependence on x), and there is a one-to-one correspondence between the section choices and the gauge transformations used in (27). In invariant language, the ghost α is given by $\alpha = w - s\ \pi\ w$, where π is the bundle projection, and w the connection.

6. Taking the lepton triplet of SU(2/1) as an example, we use the Weyl action for the massless left isodoublet $\phi(\nu_L^o\ , e_L^-)$. On the other hand, we use the Townsend (28,29) action for ψ_μ^L, an isosinglet left vector-spinor. This field thus resembles the mysterious left-handed transform of e_R^- appearing in the $\underset{\sim}{3}$ of SU(2/1), but since we are now working in the reducible and even Lie algebra $\tilde{g}$, the λ^μ_1 generator leads from $\phi(\nu_L^o, e_L^-)$ to ψ_μ^L with spin 3/2 instead of leading to a ghost (related to $\psi_\mu^L\, dx^\mu = \psi^L$). We denote by $T_{\mu\nu}$ an auxiliary Dirac spinor two-form, ψ_μ^R an auxiliary right-spinor one-form (we use 2-spinor notation)

$$L_R = \varepsilon_{\lambda\mu\nu\rho}(\overline{T^L_{\lambda\mu}}(\partial_\nu\ \psi^L_\rho + \sigma_\nu\ \psi^R_\rho) + \overline{\psi^L_\lambda}\ \sigma_\mu\ \partial_\nu\ \psi^L_\rho$$
$$+ \overline{T^R_{\lambda\mu}}\ \partial_\nu\ \psi^R_\rho) \qquad [6.1]$$

The $T_{\lambda\mu}$ equations of motion enforce the constraints

$$d\ \psi^L + \sigma \wedge \psi^R = o \quad , \qquad d\ \psi^R = o \qquad [6.2]$$

which have the solution (29)

$$\psi^L = d\ \tau^L + \sigma \wedge \nu^R$$
$$\psi^R = d\nu^R \qquad [6.3]$$

and the Lagrangian is equivalent up to the equations of motion to a Weyl Lagrangian, in terms of the right-spinor ν^R

$$L_R' \overset{o}{=} \varepsilon_{\lambda\mu\nu\rho}\ \overline{\nu^R} \wedge \sigma_\lambda \wedge \sigma_\mu \wedge \sigma_\nu \wedge \partial_\rho\ \nu^R \qquad [6.4]$$
$$= \overline{\nu^R}\ \sigma^\mu\ \partial_\mu\ \nu^R$$

The left vector-spinor ψ^L_μ in [6.1] is thus seen to represent physically a right-spinor ν^R, thus fitting the right isosinglet fermions such as $e_{\bar{R}}$. On the other hand, the formal left vector-spinor ψ^L_μ is a one-form whose vertical complement is given by

$$\hat{\psi}^L = \psi^L_\mu \, dx^\mu + \psi^L_M \, dy^M + \psi^L_{\bar{N}} \, dy^{\bar{N}}$$
$$= \psi^L_\mu \, dx + x^{L+} + x^{L-} \qquad [6.5]$$

The Bosonic spinor ghosts x^{L+} (and x^{L-}) can now be identified with the ghost state with ψ^R internal isoscalar quantum numbers (1,10,23) appearing together with the ϕ^L doublets in $\underset{\sim}{3}$ or $\underset{\sim}{4}$ of SU(2/1) (and its symmetric Curci-Ferrari extension for x^{L-} (9)).

It is indeed remarkable that the Townsend Lagrangian should thus explain both the ghost statistics and the chiral inversion in the SU(2/1) matter multiplets. We have thus found an answer to the puzzle we exposited in section 2a) of this study.

7. The interacting Lagrangian

For several reasons, there is no trivial generalization of the Abelian Lagrangian to the non Abelian case. On one hand, the Lie algebra is reducible and its Killing metric is non zero only in the A^a_μ sector, so only the Yang-Mills vector Lagrangian comes out as a natural invariant. On the other hand, if we consider as a Lagrangian for the $B^i_{\mu\nu}$ the term

$$L_B = -{}^1/12 \, (G^i_{\mu\nu\rho})^2 \qquad [7.1]$$

using the SU(2) x U(1) symmetric δ_{ij} metric, the $B^i_{\mu\nu}$ equation of the motion $D_\mu \, G_{\mu\nu\rho} = o$ enforces the constraint

$$D_\mu \, D_\nu \, G_{\mu\nu\rho} = [F_{\mu\nu} \, , \, G_{\mu\nu\rho}] = o \qquad [7.2]$$

The natural invariance of the Lagrangian under SU(2)xU(1) covariant BRS variations $SB_{\mu\nu} = D_{[\mu} \, \beta_{\nu]}$ is also lost if the constraint is not satisfied.

$$SL_B = G_{\mu\nu\rho} \, D_\mu \, D_\nu \, \beta_\rho = [G_{\mu\nu\rho} \, , \, F_{\mu\nu}] \, \beta_\rho \qquad [7.3]$$

where β_ν is the vector ghost in [5.2] and

$$S = s + [\alpha \; , \;] \qquad [7.4]$$

Inspired by Dirac's approach for Hamiltonians, we simply introduce a Lagrange multiplier K_μ whose equation of the motion enforces the differential constraints, and consider the Lagrangian

$$L'_B = -1/12 \, (\varepsilon^{\mu\nu\rho\sigma} \, (D_\mu \, (B_{\nu\rho} + D_\nu \, K_s)))^2 \qquad [7.5]$$

The equations of motion are

$$D_\mu \, (D_{[\mu}(B_{\nu\rho]} + D_\nu \, K_{\rho]})) = o$$

$$[F_{\mu\nu} \; , \; D_{[\mu} \; B'_{\nu\rho]}] = o \qquad [7.6]$$

and the system is now closed. At the same time, L'_B is invariant under the nilpotent BRS algebra (21,30,31) involving fields and ghosts from [5.2] with κ as the ghost of K_μ,

$$S\alpha = \tfrac{1}{2} \, [\alpha,\alpha]$$

$$SA_\mu = \partial_\mu \alpha$$

$$Sb = o$$

$$S\kappa \, := -b \qquad [7.7]$$

$$S\beta_\mu = D_\mu \, b$$

$$SK_\mu = - \, (\beta_\mu + D_\mu \kappa) \; := - \, \beta'_\mu$$

$$SB_{\mu\nu} = D_{[\mu} \, \beta_{\nu]} + [F_{\mu\nu} \; , \; \kappa] = D_{[\mu} \, \beta'_{\nu]}$$

Here, all the fundamental fields, i.e. those who appear in $\tilde{B}$ ($B_{\mu\nu}$, β_μ, b) have canonical dimension 1. The K_μ and its pair of ghosts have dimension 0. In a proper gauge they are expected to decouple, therefore ensuring the renormalisability of the theory by power counting, and its formal unitarity.

Note that we have answered here the puzzling features that we described in section 2b).

Our method admits a direct generalisation to the $C^a_{\mu\nu\rho}$ field, whereas $E^i_{\mu\nu\rho\;\sigma}$ has no curl in 4 dimensions. The complete classical $\tilde{g}$ gauge Lagrangian can be written, using an auxiliary 2-form $\tilde{\mathcal{L}}$ ($L_{\mu\nu}$, λ_μ , l)

$$L = -\tfrac{1}{4}\,(F_{\mu\nu}(A))^2 - 1/12\,(D_{[\mu}\,B'_{\nu\rho]})^2$$

$$- 1/48\,(D_{[\mu}\,C_{\nu\rho\sigma]} + \tfrac{1}{2}\,\{B'_{\mu\nu}\,,\,B'_{\rho\sigma}\})^2 \qquad [7.8]$$

$$B'_{\mu\nu} := B_{\mu\nu} + D_{[\mu}\,K_{\nu]} \quad , \quad C'_{\mu\nu\rho} := C_{\mu\nu\rho} + D_{[\mu}\,L_{\nu\rho]}$$

The squares are computed using the δ_{ab},δ_{ij} metric, implying $\sin^2\theta_W = .25$. The dynamics correspond to the gauging of an even lie algebra $\tilde{g}$, and the Killing metric of SU(2/1) plays no role here. We have thus answered the questions raised in section 2c).

This Lagrangian is invariant under the nilpotent BRS algebra given above for the $C_{\mu\nu\rho}$ sector:

$$S\gamma = \tfrac{1}{2}\,\{b,b\} \quad , \quad S\ell = -\gamma - \tfrac{1}{2}\{\kappa\,,b\} := -\gamma'$$

$$Sc_\mu = D_\mu\,\gamma' + \{\beta'\,,b\} \quad ,$$

$$S\lambda_\mu = -\,c_\mu - D_\mu\ell - \{K_\mu,b\} := -c'_\mu \qquad [7.9]$$

$$S\Gamma_{\mu\nu} = D_{[\mu}\,c'_{\nu]} + \tfrac{1}{2}\,\{\beta'_{[\mu},\beta'_{\nu]} + \{B'_{\mu\nu},b\}$$

$$SL_{\mu\nu} = -\Gamma_{\mu\nu} - D_{[\mu}\,\lambda_{\nu]} - \{B'_{\mu\nu},\kappa\} - \tfrac{1}{2}\{\beta'_{[\mu},K_{\nu]} := -\Gamma'_{\mu\nu}$$

$$SC_{\mu\nu\rho} = D_{[\mu}\,\Gamma'_{\nu\rho]} + \{B'_{[\mu\nu},\beta'_{\rho]}$$

At the linearized level, we recover the Weinberg-Salam spectrum.

Note that the closure of the BRS algebra is equivalent, through complexification of the y and projection in the dy, dy sectors to the closure of the extended Curci-Ferrari algebra. No additional calculations are required.

8. Let us now construct a BRS algebra (21,31) for chiral spinor fields. We start from a left chiral spin 1/2 R^+ multiplet ϕ^L and write:

$$S\phi^L = o \qquad [8.1]$$

We now introduce a left chiral vector spinor ψ^L_μ, its left chiral Bose spin 1/2 ghost x^{L+} and an auxiliary spin 1/2 Fermion η^L, all valued in R^-. The nilpotent BRS algebra is uniquely defined as

$$
\begin{aligned}
S\, x^+ &= -\{b\,,\,\phi\} \\
S\, \eta &= x^+ + \{\kappa\,,\,\phi] \\
S\, \psi_\mu &= D_\mu\, x^+ + \{\beta_\mu\,,\,\phi\} + \{\kappa\,,\,D_\mu\phi\}
\end{aligned}
\qquad [8.2]
$$

We can now construct the covariant differential

$$
\begin{aligned}
\tilde{D}_{[\mu}\,\psi_{\nu]} &= D_{[\mu}\,\psi_{\nu]} + \{B_{\mu\nu}\,,\phi\} + \{K_{[\mu}\,,\,D_{\nu]}\phi\} \\
&\quad + [F_{\mu\nu}\,,\eta] \\
S\,\tilde{D}_\psi &= o
\end{aligned}
\qquad [8.3]
$$

and the invariant spinor

$$
\begin{aligned}
\tilde{\psi}_\mu &= \psi_\mu + \{K_\mu\,,\,\phi\} + D_\mu \eta \\
S\tilde{\psi}_\mu &= o
\end{aligned}
\qquad [8.4]
$$

Townsend's Lagrangian now generalizes into:

$$
\begin{aligned}
L_{\frac{1}{2}} &= \varepsilon^{\lambda\mu\nu\rho}\,(\overline{\Gamma^L_{\mu\nu}}\,(\tilde{D}_\nu\,\psi^L_\rho + \sigma_\nu\,\tilde{\psi}^R_{\hat{\rho}}) \\
&\quad + \overline{\Gamma^R_{\lambda\mu}}\,\tilde{D}_\nu\,\psi^R_\rho \\
&= \overline{\psi^L_\lambda}\,\sigma_\mu\,\tilde{D}_\nu\,\psi^L_\rho) + \overline{\phi^L}\,\not{D}\,\phi^L
\end{aligned}
\qquad [8.5]
$$

At the linearized level, we recover Townsend's Lagrangian and propagte a left doublet $\phi^L(\nu^o_L\,,\,\varepsilon^-_L)$ and a right singlet $\phi^R(e^-_R)$. The dynamical part of the Lagrangian is the generalized Rarita-Schwinger term containing a coupling $\psi^L_\lambda\,\sigma_\mu\,\{B_{\nu\rho},\phi^R\}$ which replaces the usual mass term of the Weinberg-Salam model.

9. The pattern of spontaneous symmetry breakdown is crucial for the application of the model to experiment. Obviously, the tensor $B^i_{\mu\nu}$ may not take a vacuum expectation value without

breaking at the same time the Lorentz group. However, its central scalar Bose ghost $h^i = B^i{}_{\overline{MN}}\, dy^M \wedge dy^{\overline{N}}$ has the quantum numbers of the ordinary Higgs field. Furthermore the term $(DC)^2$ contains $(B^i_{\mu\nu})^4$, which generalizes in the ghost expansion to $(h^i)^4$. We therefore conjecture that, in the fully quantized theory $\langle h^i \rangle_o \neq o$. We remark that a term $S\overline{S}\ (\overline{\psi}.\sigma\eta) = \overline{\psi}.\sigma\ \{h\ ,\phi\} + \ldots$ is BRS admissible and represents an arbitrary (electron) mass term. Spontaneous symmetry breakdown seems also deeply interwined with the problem of the differential constraints in this theory. Indeed, the theory is unstable for $g \to o$ as the constraint $[F\ , *G] \sim [dA\ , *dB] = o$ is present for $g \neq 0$ but is of order 0 in g. The constraint obviously admits as a solution the usual symmetry breaking pattern, G^i in the group direction μ_o , F^θ confined to the photon-direction $1/\sqrt{2}\ (\mu_3 + \sqrt{3}\ \mu_8)$

Clearly, we still have to investigate the manner in which the spontaneous symmetry breakdown should be implementated.

References

1. Ne'eman, Y. (1979) Phys. Lett. B81, pp. 190-194.
2. Fairlie, D. B. (1979) Phys. Lett. B82, pp. 97-100.
3. Ne'eman, Y. & Sternberg, S. (1980) Proc. Nat. Acad. Sc., USA, 77, pp. 3127-3131.
4. Ne'eman, Y. & Thierry-Mieg, J. (1982) Phys. Letter. B108, pp. 399-402.
5. Ne'eman, Y., Sternberg, S. & Thierry-Mieg, J. (1982) Proc. Rome (GUD) Int. Conf., G. Ciapetti et al eds., Frascati Lab. pub. pp. 89-92.
6. Becchi, C., Rouet, A. & Stora, R. (1975) Comm. Math. Phys. 42, 126.
7. Curci, G. & Ferrari, F. (1975) Nuovo Cimento 30A, pp. 155-168.
8. Ne'eman, Y. (1980) in Unific ation of the Fundamental Interactions, ed. S. Ferrara et al, Plenum Press (New York, N.Y.) pp. 89-94.
9. Thierry-Mieg, J. & Ne'eman, Y. (1982) Nuovo Cimento 71A, 104.
10. Ne'eman, Y. & Thierry-Mieg, J. (1980) in Differential Geometrical Methods in Mathematical Physics (Proc. Aix en Provence & Salamanca 1979); P. L. Garcia, A. Perez Rendon & J. M. Souriau eds., Springer Verlag Lecture Notes in Math. 836, (Berlin/Heidelberg/N.Y.), pp. 318-348.
11. Thierry-Mieg, J. & Ne'eman, Y. (1982) Proc. Nat. Acad. Sci. USA 79, pp. 7068-7072.
12. Cartan, E. C. R. Acad. Sc. Paris (1926) 182, pp. 956-958.
13. Kalb, M. & Ramond, P. (1974) Phys. Rev. D9, pp. 2273-2285.
14. Cremmer, E. & Scherk, J. (1974) Nucl. Phys. B72, pp. 117-124.
15. Cremmer, E. & Julia, B. (1979) Nucl. Phys. B159, pp. 141-213.
16. Thierry-Mieg, J. (1979) Harvard Preprint HUMTP B86, unpub.
17. Baulieu, L. & Thierry-Mieg, J. (1982) Nucl. Phys. B197, pp. 477-508.

18. Thierry-Mieg, J. (1978) Thèse. de Doctorat d'Etat (Paris-Sud)
19. Ne'eman, Y. (1979) in Proceedings XIXth International Conf. on High Energy Physics, Tokyo 1978, S. Hommer, M. Kawaguchi and H. Miyazawa eds., Phys. Soc. Jap., Tokyo 1979,pp.552-554.
20. Thierry-Mieg, J. & Ne'eman, Y. (1979) Ann. Phys. (N.Y.) 123, p. 247.
21. Thierry-Mieg, J. (1980) Jour. Math. Phys. 21, pp.2834-2838.
22. Thierry-Mieg, J. (1980) Nuovo Cim. 56A, p. 396.
23. Ne'eman, Y. & Thierry-Mieg, J. (1980) Proc. Nat. Acad. Sc., USA, 77, pp. 720-723.
24. Quiros, M., de Urries, F. J., Hoyos, J., Mazon, M. L. & Rodriguez, E. (1981) J. Math. Phys. 22, p. 1767.
25. Bonora, L. & Tonin, M. (1981) Phys. Lett. 98B, pp. 48-50.
26. Delbourgo, R. & Jarvis, P. D. (1982) J. Phys. A. Math. Gen. 15, p. 611.
27. Leinaas, J. M. & Olaussen, K. (1982) Phys. Lett. 108B, p.199.
28. Townsend, P. K. (1980) Phys. Lett. B90, pp. 275-276.
29. Deser, S., Townsend, P. K. & Siegel, W. (1981) Nucl. Phys. B184, pp. 333-350.
30. Freedman, D. Z. (1977) Caltech rep. 68-624, unpub.
31. Freedman, D. Z. & Townsend, P. K. (1981) Nucl. Phys. B177. pp. 282-296.

CASIMIR ELEMENTS OF LIE SUPERALGEBRAS

M. Scheunert*

Physikalisches Institut der Universität Bonn

The Casimir elements of the general linear Lie superalgebras are investigated. For a standard sequence of Casimir elements, the eigenvalues in a highest weight module are calculated. The results are then used to get information about the structure of the algebra of all Casimir elements.

1. INTRODUCTION

In the following I am going to discuss the Casimir elements of Lie superalgebras[1,2]. I do not want to argue that this should be an interesting topic: The Casimir elements play an important role in the representation theory of semi-simple Lie algebras, and we expect this will also be true for Lie superalgebras. This is indeed the case, as is evident, for example, from the following papers by V. G. Kac[3,4], which are sort of a basis for my present talk.

At the same time, these papers clearly exhibit some peculiarities of the Lie superalgebra case. Recall that under a (finite-dimensional) irreducible representation a Casimir element is represented by a scalar multiple of the identity. In the case of semi-simple Lie algebras, the scalar factors, i.e. the eigenvalues of the Casimir operators, completely specify the representation: For any two non-equivalent irreducible representations there exists a Casimir element whose corresponding eigenvalues are different. We say that the Casimir elements *separate* the irreducible representations. For simple Lie superalgebras the situation is more complicated (even if we restrict our attention to the so-called basic classical ones). Typically, that is, for irreducible representations which are "in general position", the foregoing

S. Sternberg (ed.), Differential Geometric Methods in Mathematical Physics, 115–124.

statement still holds true; however, certain special irreducible representations, the so-called non-typical ones, are no longer separated by the Casimir elements. This fact is at the heart of a second well-known pathology: A (finite-dimensional) representation of a simple Lie superalgebra is not necessarily completely reducible.

My talk is mainly devoted to the calculation of the eigenvalues of a specified sequence of Casimir elements. The results will clearly show how the Casimir elements fail to separate the irreducible representations. They will also shed some light on the structure of the algebra of all Casimir elements.

Of course, the eigenvalues of the quadratic Casimir elements are well-known. Some special higher order cases have been considered by A. B. Balantekin and I. Bars [5,6] and by P. D. Jarvis and M. K. Murray [7]. Apart from this only very little can be found in the present literature.

To reduce the technicalities to a minimum I shall mainly consider the general linear Lie superalgebras and only briefly comment on the question of how the results generalize to the special linear and the orthosymplectic ones.

2. CASIMIR ELEMENTS OF THE GENERAL LINEAR LIE SUPERALGEBRAS

Let us first fix the notation. In the following, the base field may be any commutative field of characteristic zero. Consider a finite-dimensional Z_2-*graded vector space*

$$V = V_{\bar{0}} \oplus V_{\bar{1}} .$$

Choose a *homogeneous basis* $(e_i)_{i \in I}$ of V, let $\eta_i \in Z_2$ denote the degree of e_i, and set

$$\sigma_i = (-1)^{\eta_i} , \qquad d = \sum_{i \in I} \sigma_i = \dim V_{\bar{0}} - \dim V_{\bar{1}} .$$

By means of the *super-commutator* $\langle \; , \; \rangle$, the vector space of all linear mappings of V into itself is converted into the *general linear Lie superalgebra* $gl(V_{\bar{0}}, V_{\bar{1}})$. Let E_{ij} ; $i,j \in I$, be the linear mapping of V into itself whose matrix with respect to (e_i) is given by

$$(E_{ij})_{k\ell} = \delta_{ik}\delta_{j\ell} .$$

Of course, the E_{ij} form a basis of $gl(V_{\bar{0}}, V_{\bar{1}})$. We note that E_{ij} is homogeneous of degree $\eta_i + \eta_j$ and that

$$\langle E_{ij}, E_{k\ell} \rangle = \delta_{jk} E_{i\ell} - (-1)^{(\eta_i+\eta_j)(\eta_k+\eta_\ell)} \delta_{\ell i} E_{kj} .$$

Let h denote the subspace of $gl(V_{\bar{0}},V_{\bar{1}})$ which is generated by the elements E_{ii} ; $i \in I$. Of course, h is a *Cartan subalgebra* of $gl(V_{\bar{0}},V_{\bar{1}})$, it consists of all linear mappings whose matrix with respect to the basis (e_i) is diagonal.

The corresponding *root space decomposition* of $gl(V_{\bar{0}},V_{\bar{1}})$ is easily constructed. For every $j \in I$ we define a linear form ε_j on h through the equation

$$He_j = \varepsilon_j(H)e_j \quad \text{for all} \quad H \in h ,$$

that is

$$\varepsilon_j(E_{ii}) = \delta_{ij} \quad \text{for all} \quad i,j \in I .$$

Then we have

$$\langle H,E_{ij} \rangle = (\varepsilon_i - \varepsilon_j)(H)E_{ij}$$

for all $H \in h$ and all $i,j \in I$. Consequently,

$$\Delta = \{\varepsilon_i - \varepsilon_j \mid i,j \in I \; ; \; i \neq j\}$$

is the *root system* of $gl(V_{\bar{0}},V_{\bar{1}})$ with respect to h and E_{ij} is a *root vector* corresponding to $\varepsilon_i - \varepsilon_j$; $i \neq j$. The root $\varepsilon_i - \varepsilon_j$ is called *even/odd* depending on whether E_{ij} is even/odd, that is, if $\sigma_i \sigma_j = \pm 1$.

All this looks the same as in the case of the general linear Lie algebra. Of course, this is not a miracle, since the adjoint action of h is the same in both cases. A crucial difference arises if we introduce an adequate bilinear metric on the dual h^* of h . The supertrace Str yields an invariant, super-symmetric, non-degenerate bilinear form on $gl(V_{\bar{0}},V_{\bar{1}})$,

$$(A,B) \longrightarrow \mathrm{Str}(AB) ,$$

and this form induces the following symmetric, non-degenerate *bilinear form* $(\mid)$ *on* h^*

$$(\varepsilon_i \mid \varepsilon_j) = \sigma_i \delta_{ij} \quad \text{for all} \quad i,j \in I .$$

Note that the odd roots are all isotropic with respect to $(\mid)$.

Let us next specify a system of positive roots or, what amounts to the same, a basis of the root system Δ. Choose any total ordering on I and let $i_1,i_2,\ldots,i_p$ be the strictly increasing sequence of elements of I. We set

$$\alpha_q = \varepsilon_{i_q} - \varepsilon_{i_{q+1}} \quad \text{for} \quad 1 \le q \le p-1 .$$

The α_q are linearly independent, and any root is a linear combination of them with integral coefficients which are either all positive or all negative. Thus $(\alpha_q)_{1\leq q\leq p-1}$ is a *basis of* Δ in the usual sense and the corresponding *system of positive roots* is

$$\Delta^+ = \{\varepsilon_i - \varepsilon_j \,|\, i,j\in I \; ; \; i<j\} \; .$$

Later on we shall need the element $\rho \in h^*$ which is defined to be half the sum of the even positive roots minus half the sum of the odd positive roots. It is easy to check that

$$\rho = \sum_{i\in I} \sigma_i r_i \varepsilon_i$$

with

$$r_i = \sigma_i \rho(E_{ii}) = \frac{1}{2}\Big(\sum_{j>i}\sigma_j - \sum_{j<i}\sigma_j\Big) .$$

We have for all integers $p \geq 0$

$$\sum_{i\in I} \sigma_i r_i^{2p+1} = 0$$

$$\sum_{q=0}^{p} \binom{2p+1}{2q} \sum_{i\in I} \sigma_i (2r_i)^{2q} = d^{2p+1} \; .$$

Let us now introduce the concept of a *highest weight module* over $gl(V_{\bar{0}},V_{\bar{1}})$. By definition, this is a (possibly infinite-dimensional) Z_2-graded vector space W, endowed with a graded representation of $gl(V_{\bar{0}},V_{\bar{1}})$, such that the following conditions are satisfied. There exist a homogeneous non-zero element $w\in W$ and a linear form $\Lambda \in h^*$ such that

I) w generates W as a $gl(V_{\bar{0}},V_{\bar{1}})$-module.

II) w is annihilated by all E_{ij} with $i<j$ (that is, by the root vectors corresponding to the positive roots).

III) $Hw = \Lambda(H)w$ for all $H\in h$.

The linear form Λ is uniquely fixed, the vector w is determined up to a scalar factor. We call Λ the *highest weight* and w a *generator* of W .

One can show that for any linear form $\Lambda\in h^*$ there exist (possibly several inequivalent) highest weight modules with highest weight Λ . All finite-dimensional irreducible modules are highest weight modules, provided the base field is algebraically closed.

We are now ready to discuss the Casimir elements. Let U denote the *universal enveloping algebra* of $gl(V_{\bar{0}},V_{\bar{1}})$. Recall that U is an associative Z_2-graded algebra with unit element, that $gl(V_{\bar{0}},V_{\bar{1}})$ is canonically embedded in U, that 1 and $gl(V_{\bar{0}},V_{\bar{1}})$ generate the algebra U, and that any representation of $gl(V_{\bar{0}},V_{\bar{1}})$ can be extended to a representation of U. Moreover, a generalized version of the Poincaré, Birkhoff, Witt theorem holds true.

We consider the *center* Z (in the super-sense) of U. By definition, an element $C \in U$ belongs to Z if and only if it super-commutes with all elements of U or, equivalently, if it super-commutes with all elements of $gl(V_{\bar{0}},V_{\bar{1}})$,

$$\langle C,A\rangle = 0 \quad \text{for all} \quad A \in gl(V_{\bar{0}},V_{\bar{1}}) .$$

The elements of Z are called (generalized) *Casimir elements* of $gl(V_{\bar{0}},V_{\bar{1}})$. It can be shown that all Casimir elements are even and hence commute with all elements of U.

A standard construction of Casimir elements goes as follows. Consider a matrix $(X_{ij})_{i,j\in I}$ with elements from U. We call this matrix *tensorial* if it satisfies the following conditions:

I) X_{ij} is homogeneous of degree $\eta_i+\eta_j$.

II) $\langle E_{ij},X_{k\ell}\rangle = \delta_{jk}X_{i\ell} - (-1)^{(\eta_i+\eta_j)(\eta_k+\eta_\ell)}\delta_{\ell i}X_{kj}$.

Note that (E_{ij}) is tensorial. It is easy to show that:

1. If (X_{ij}) and (X'_{ij}) are tensorial, then so is the matrix (Y_{ij}) defined by

$$Y_{ij} = \sum_{k\in I} \sigma_k X_{ik} X'_{kj} .$$

2. If (X_{ij}) is tensorial, then $\sum_{i\in I} X_{ii}$ is a Casimir element.

Consequently,

$$C_r = \sum_{i_1,\dots,i_r} \sigma_{i_1}\sigma_{i_2}\cdots\sigma_{i_{r-1}} E_{i_r i_1} E_{i_1 i_2} \cdots E_{i_{r-1} i_r}$$

is a Casimir element, for all integers $r \geq 1$ (note that the product on the right is calculated in U).

Now let $\Lambda \in h^*$ and let W be a highest weight module with highest weight Λ. It is easy to see that any Casimir element C is represented in W by a scalar multiple of the identity. The scalar factor depends only on Λ and C, it will be denoted by

$\chi_\Lambda(C)$. One shows:

For fixed $\Lambda \in h^*$, the mapping $C \longrightarrow \chi_\Lambda(C)$ is an *algebra homomorphism* of Z onto the base field.

For fixed $C \in Z$, the mapping $\Lambda \longrightarrow \chi_\Lambda(C)$ is a *polynomial function* on h^* which *uniquely* determines the Casimir element C.

As in the case of semi-simple Lie algebras, it will be advantageous to consider $\chi_\Lambda(C)$ as a function of $\lambda = \Lambda+\rho$.

Our next task is to calculate $\chi_\Lambda(C_r)$ for all $\Lambda \in h^*$ and all integers $r \geq 1$. We obtain

$$\chi_\Lambda(C_r) = c_r(\Lambda+\rho)$$

$$c_r(\lambda) = \sum_{i,j\in I} (A^r(\lambda))_{ij}\sigma_j ,$$

where the I×I-matrix $A(\lambda)$ is given by

$$A_{ij}(\lambda) = (\sigma_i\lambda(E_{ii}) + \frac{1}{2}(d-\sigma_i))\delta_{ij} - \sigma_i\xi_{ij} ,$$

with

$$\xi_{ij} = \begin{cases} 1 & \text{if } i<j \\ 0 & \text{if } i\geq j . \end{cases}$$

To evaluate the functions c_r we follow the classical approach by A. M. Perelomov and V. S. Popov [8]. Let z be an indeterminate. Introduce the *generating function* for the c_r by

$$G(z) = \sum_{r\geq 0} c_r z^r .$$

Then

$$G(z) = \frac{1}{z}(1 - e^{-f(z)}) + de^{-f(z)}$$

with

$$f(z) = \sum_{r\geq 2}\left(\sum_{s\geq 0} \frac{1}{r}\binom{r}{r-1-2s} 2^{-2s} R_{r-1-2s}\right) z^r$$

$$R_p = \sum_{m=0}^{p} \binom{p}{m}\left(\frac{d}{2}\right)^{p-m} Q_m$$

and where the polynomial functions Q_m of λ are defined by

$$Q_m = \sum_{i\in I}\sigma_i(\ell_i^m - r_i^m)$$

$$\ell_i = \sigma_i \lambda(E_{ii}) , \qquad r_i = \sigma_i \rho(E_{ii}) .$$

The main conclusion that we draw from these formulae is the following.

Any c_r *can be written as a polynomial in the* Q_m *, and* $c_r - Q_r$ *has degree* $\leq r-1$ *, for all integers* $r \geq 1$ *. Consequently, any* Q_m *can also be written as a polynomial in the* c_r *.*

For small values of r , the c_r are easily calculated:

$$c_0 = d , \quad c_1 = Q_1 , \quad c_2 = Q_2$$

$$c_3 = Q_3 + \frac{1}{2} d\, Q_2 - \frac{1}{2} Q_1^2 - \frac{1}{4}(d^2-1)\, Q_1$$

$$c_4 = Q_4 + d\, Q_3 - Q_1 Q_2 + \frac{1}{2} Q_2 - \frac{1}{2} d\, Q_1^2 - \frac{1}{4} d\,(d^2-1)\, Q_1 .$$

Next we note that the functions c_r have the following basic properties (because the Q_m do):

1. The c_r are invariant under the transformation $\ell_i \longrightarrow \ell_{\pi(i)}$, where π is any permutation of I which satisfies

$$\sigma_{\pi(i)} = \sigma_i \quad \text{for all } i \in I .$$

This is nothing but the invariance of c_r under the Weyl group.

2. Choose $i,j \in I$ with $\sigma_i \sigma_j = -1$. If we set $\ell_i = \ell_j$, then c_r is independent of this variable.

More generally, V. G. Kac has shown [3,4]:

Let C *be any Casimir element. Define the polynomial function* c *on* h^* *by*

$$c(\lambda) = \chi_{\lambda-\rho}(C) \quad \text{for all } \lambda \in h^* .$$

Then c *has the properties 1. and 2. above,* and the proof reveals that 2. has its origin in the isotropy of the odd roots.

Actually, V. G. Kac has proved this under the assumption that the ordering on I is chosen such that

$$i < j \quad \text{if} \quad \sigma_i = 1 \quad \text{and} \quad \sigma_j = -1$$

(the indices corresponding to even basis elements come first). This choice has the advantage that our basis of the root system contains at most one odd root. Let us, therefore, *assume that this condition is fulfilled.*

Now we note that the non-typical representations are exactly those finite-dimensional irreducible representations whose highest weight $\Lambda = \lambda - \rho$ satisfies

$$\ell_i = \ell_j \quad \text{for some } i,j \in I \quad \text{with } \sigma_i \sigma_j = -1 .$$

Thus property 2. is the reason why the Casimir elements do not separate the non-typical representations.

At this point I would like to mention that everything that I have said thus far can also be carried through for the special linear (with $d \neq 0$) and the orthosymplectic Lie superalgebras [9]. In particular, the Casimir elements C_r can be constructed and the generating function for their eigenvalues can be evaluated.

3. SUPERSYMMETRIC POLYNOMIALS

In the final part of my talk I would like to comment on what our results imply for the structure of the algebra of all Casimir elements. This topic is still under investigation, hence only partial results can be presented.

We have already noted that the mapping which associates with any Casimir element C the polynomial function c on h^* defined by

$$c(\lambda) = \chi_{\lambda-\rho}(C) \quad \text{for all } \lambda \in h^* ,$$

is an *injective algebra homomorphism*. Thus, instead of considering the algebra Z of all Casimir elements C, we may equally well investigate the algebra of the corresponding polynomial functions c.

Let us first adjust our notation. We set

$$\dim V_{\bar{0}} = n , \qquad \dim V_{\bar{1}} = m .$$

Choose n indeterminates $x_1, \ldots, x_n$ and m indeterminates $y_1, \ldots, y_m$. For every integer $r \geq 1$, define the polynomial P_r in the x_i and y_j by

$$P_r = \sum_{i=1}^{n} x_i^r - \sum_{j=1}^{m} y_j^r .$$

Let $S(n,m)$ denote the algebra of polynomials generated by 1 and the P_r ; $r \geq 1$. Under the obvious substitution of variables, this algebra corresponds isomorphically to the subalgebra Z' of Z which is generated by 1 and the C_r ; $r \geq 1$.

The polynomials P_r generalize the classical power sums to the super-case and hence may be called the supersymmetric power sums. Similarly, the elementary symmetric polynomials and the complete symmetric polynomials can be generalized [5]. It is easy to see that both the elementary supersymmetric polynomials and the complete supersymmetric polynomials also generate the algebra $S(n,m)$. Thus it seems reasonable to call the polynomials in $S(n,m)$ *supersymmetric*.

Consider next the algebra $T(n,m)$ consisting of all polynomials Q in the x_i and y_j, which satisfy the following conditions:

1. Q is symmetric in the x_i and in the y_j.

2. Let $i \in \{1,\dots,n\}$ and $j \in \{1,\dots,m\}$. If we set $x_i = y_j$ in Q, the resulting polynomial is independent of this variable.

Our foregoing discussion shows that the algebra of polynomials corresponding to Z is a subalgebra of $T(n,m)$ containing $S(n,m)$.

I *conjecture* that for all n,m

$$S(n,m) = T(n,m) .$$

Stated differently: Any polynomial Q in the x_i and y_j which satisfies the conditions 1. and 2. above can be written as a polynomial in the P_r; $r \geq 1$.

Note that *this conjecture implies that* $Z' = Z$, *i.e., that* 1 *and the* C_r; $r \geq 1$, *generate the algebra of all Casimir elements.*

The following partial results support the conjecture:
If $n = 0$ or $m = 0$, the result is classical.
The conjecture is true if $n = 1$ or $m = 1$.
The conjecture holds for polynomials of degree ≤ 10.
Let e_r; $0 \leq r \leq n$, denote the elementary symmetric polynomials in the x_i and let

$$D = \prod_{i,j} (x_i - y_j) .$$

Then the polynomials $e_r D$; $0 \leq r \leq n$, are in $S(n,m)$. Of course, the analogous result holds for the variables y_j. This implies that for any polynomial R in the x_i and y_j, which is symmetric in the x_i and in the y_j, there exists an integer $q \geq 0$ such that $D^q R$ is in $S(n,m)$. This result is due to V. G. Kac [3,4]. Obviously, DR is in $T(n,m)$.

Irrespective of whether our conjecture holds or not, we can prove the following facts (which one should contrast with the corresponding classical results).

The algebra $S(1,1) = T(1,1)$ consists of all polynomials of the form

$$\gamma + (x_1 - y_1) R(x_1, y_1)$$

with a constant γ and an arbitrary polynomial R in x_1, y_1.

Note that this algebra is not finitely generated. Consequently, *for* $n,m \geq 1$ *none of the algebras* $S(n,m)$, $T(n,m)$, Z', Z *is finitely generated.*

On the other hand, consider four families $\xi = (\xi_i)_{1 \leq i \leq n}$, $\eta = (\eta_j)_{1 \leq j \leq m}$, $\xi' = (\xi'_i)_{1 \leq i \leq n}$, and $\eta' = (\eta'_j)_{1 \leq j \leq m}$ of elements from the base field. *If*

$$P_r(\xi,\eta) = P_r(\xi',\eta') \quad \text{for} \quad 1 \leq r \leq n+m ,$$

then

$$Q(\xi,\eta) = Q(\xi',\eta') \quad \text{for all} \quad Q \in T(n,m) .$$

Transcribed to the Casimir elements, this implies that *concerning the separation of the highest weight modules,* i.e., concerning the labeling problem, *the Casimir elements* C_r ; $1 \leq r \leq n+m$, *yield as much as the algebra of all Casimir elements does.*

Let us close our discussion by the remark that the supersymmetric polynomials are equally relevant to the special linear and the orthosymplectic Lie superalgebras.

*Supported by the Deutsche Forschungsgemeinschaft

REFERENCES

1 V. G. Kac 1977, Adv. Math. 26, pp. 8 - 96

2 M. Scheunert, *The theory of Lie superalgebras.* Lecture Notes in Mathematics 716; Springer, Berlin, Heidelberg, New York 1979

3 V. G. Kac 1977, Comm. Alg. 5, pp. 889 - 897

4 V. G. Kac in *Differential geometrical methods in mathematical physics II*, Bonn 1977. Lecture Notes in Mathematics 676; Springer, Berlin, Heidelberg, New York 1978, pp. 597 - 626

5 A. B. Balantekin and I. Bars 1981, J. Math. Phys. 22, pp. 1149 - 1162

6 A. B. Balantekin 1982, J. Math. Phys. 23, pp. 486 - 489

7 P. D. Jarvis and M. K. Murray, *Casimir invariants, characteristic identities and tensor operators for "strange" superalgebras.* Univ. of Tasmania preprint, December 1981

8 A. M. Perelomov and V. S. Popov 1968, Izv. Akad. Nauk SSSR, Ser. Mat. 32. Engl. transl. Math. USSR, Izv. 2, pp. 1313 - 1335

9 M. Scheunert, *Eigenvalues of Casimir operators for the general linear, the special linear, and the orthosymplectic Lie superalgebras.* Univ. of Bonn preprint HE-82-26, October 1982

NORMAL FORM FOR HAMILTONIAN VECTORFIELDS WITH PERIODIC FLOW

Richard Cushman

Mathematics Institute, Rijksuniversiteit Utrecht,
The Netherlands

Abstract:

Suppose that f_0 is a Hamiltonian function on a symplectic manifold (M,Ω) such that its Hamiltonian vectorfield has only periodic orbits. Then the smooth formal power series Hamiltonian f_ε with zeroth order term f_0 can be brought into normal form with respect to f_0, that is, there is a smooth formal power series symplectic diffeomorphism ϕ_ε of (M,Ω) such that the Poisson bracket of $\phi_\varepsilon{}^*f_\varepsilon$ and f_0 vanishes on M. In the special case where f_0 is the Hamiltonian for the geodesic flow on the tangent bundle to the n-sphere, it is shown that f_ε is in normal form if and only if every term of the power series of f_ε is a smooth function of the components of the $SO(n+1,\mathbb{R})$-momentum mapping associated to the orthogonal symmetries of f_0.

0. INTRODUCTION.

On the symplectic manifold (M,Ω) consider a one parameter family of smooth Hamiltonian functions $f_\varepsilon = \sum_{n\geq 0} \frac{\varepsilon^n}{n!} f_n$ which is a formal power series in ε. Suppose that the Hamiltonian vectorfield X_{f_0} of the zeroth order term has only periodic integral

S. Sternberg (ed.), Differential Geometric Methods in Mathematical Physics, 125–144.

curves. Then following Deprit (1982) we say that f is in normal form with respect to f_0 if and only if the Poisson bracket of f_ε and f_0 vanishes identically on M.

In the first section we show that the vector space of smooth functions on M is the topological direct sum of the kernel and image of the Lie derivative of X_{f_0}. Thus, using Lie series (Deprit 1969) f_ε may be brought into normal form.

The remaining sections are concerned with the special case of Hamiltonian perturbations K_ε of the geodesic vectorfield on the n-sphere. Here $M = TS^n$, Ω is the restriction of the standard symplectic form ω, on the tangent bundle $T\mathbb{R}^{n+1}$ of $\mathbb{R}^{n+1}$, to the tangent bundle TS^n of S^n, and the zeroth order Hamiltonian K_0 is the length in $\mathbb{R}^{n+1}$ of a tangent vector to S^n. Clearly all the integral curves of X_{K_0} (with project to geodesics on S^n) are periodic. The principal result in the second section is that the terms in the normal form of K with respect to K_0 are smooth functions of the components of the $SO(n+1,\mathbb{R})$ momentum mapping coming from the linear $SO(n+1,\mathbb{R})$ action on $S^n \subset \mathbb{R}^{n+1}$. The case n = 3 is of special interest. Using the relation between the geodesic vectorfield on S^3 and the Kepler problem in $\mathbb{R}^3$ (Moser 1970), we show that any smooth function, which is constant on the bounded orbits of the Kepler problem in $\mathbb{R}^3$, is a smooth function, of the energy, and the components of the angular momentum and the eccentric axis (Pollard 1966). This result is a folk theorem in celestial mechanics whose published proof the author could not find.

In the third section invariant theory is used to carry out the reduction process (Marsden & Weinstein 1974) for axially symmetric Hamiltonian perturbations of the geodesic vectorfield on S^3. A worked out application of the technique of reduction to the problem of the critical inclination may be found in Cushman (1982).

1. NORMAL FORM: GENERAL THEORY

We begin with the definition of normal form. On the symplectic manifold (M,Ω) the space of smooth functions $C^{\infty}(M)$ is a Poisson algebra under pointwise multiplication . and Poisson bracket $\{,\}$, that is, for every $f, g, h \in C^{\infty}(M)$

$$\{f,\{g,h\}\} = \{\{f,g\},h\} + \{g,\{f,h\}\}$$

and

$$\{f,g.h\} = \{f,g\}.h + g.\{f,h\}.$$

Let $A = \{f_{\varepsilon} = \sum_{n\geqslant 0} \frac{\varepsilon^n}{n!} f_n \in C^{\infty}(M) \mid f_n \in C^{\infty}(M)\}$. Then A is the graded Poisson algebra of smooth formal power series in ε. The grading on A is determined by powers of ε. Let $f_{\varepsilon} \in A$, then f_{ε} is in normal form with respect to f_0 if and only if

$$L_{f_0} f_{\varepsilon} = \{f_{\varepsilon}, f_0\} = 0 \tag{1}$$

where L_{f_0} is the Lie derivative with respect to X_{f_0}. Note that (1) is equivalent to

$$\{f_0, f_n\} = 0 \text{ for all } n \geqslant 0.$$

Suppose that for a given $f_0 \in C^{\infty}(M)$ we have

$$C^{\infty}(M) = \ker L_{f_0} \oplus \operatorname{im} L_{f_0}. \tag{2}$$

Then the following standard argument (Abraham & Marsden 1978) shows how to bring f_{ε} into normal form with respect to f_0. Let $g \in C^{\infty}(M)$ and suppose that m are the co-ordinates of $m \in M$ in a Darboux chart. The Lie series

$$\phi_g(m) = (\exp \varepsilon L_g)m = \sum_{n\geqslant 0} \frac{\varepsilon^n}{n!} L_g^n m \tag{3}$$

defines a global formal symplectic diffeomorphism of (M,Ω)

because ϕ_g is the time ε mapping of the flow of the Hamiltonian vectorfield X_g. Under the symplectic change of co-ordinates ϕ_g, f_ε becomes

$$F_\varepsilon = \phi_g^* f_\varepsilon = f_\varepsilon o \phi_g = \sum_{n \geqslant 0} \frac{\varepsilon^n}{n!} L_g^n f_\varepsilon \tag{4}$$

$$= f_0 + \varepsilon(L_g f_0 + f_1) + \frac{\varepsilon^2}{2!}(L_g^2 f_0 + 2L_g f_1 + f_2) + \ldots . \tag{5}$$

To bring f_ε into normal form to first order in ε we use (5) to determine g and hence by (3) the symplectic diffeomorphism ϕ_g. By hypothesis (2), the first order terms in F_ε may be written

$$L_g f_0 + f_1 = -L_{f_0} g + f_1' + f_1'' \tag{6}$$

where

$$f_1' \in \text{Ker } L_{f_0} \text{ and } f_1'' \in \text{im } L_{f_0} .$$

Since $f_1'' \in \text{im } L_{f_0}$

$$L_{f_0} g = f_1''$$

may be solved for g. With this choice of g, the first order term in F is f_1'. Thus F_ε is in normal form with respect to f_0 to first order in ε. To bring F_ε into normal form to second order we apply the symplectic co-ordinate change $\phi_{\varepsilon h}$. Using (4) we see that $G_\varepsilon = \phi_{\varepsilon h}^* F_\varepsilon$ has the same first order terms as F_ε. The second order terms of G_ε are

$$L_h f_0 + L_g^2 f_0 + 2L_g f_1 + f_2$$

which by (2) may be solved for h as in (6). Repeating this argument gives a formal symplectic diffeomorphism which brings f_ε into normal form with respect to f_0 to any order in ε.

Thus the question of normal form is reduced to the problem of finding those smooth functions f_0 on M such that L_{f_0} splits $C^\infty(M)$ as in (2). In the classical case, where (M,Ω) is (TR^{n+1},ω)

and $f_\varepsilon = \sum_{n\geq 0} \frac{\varepsilon^n}{(n+2)!} f_n$ with f_n being a homogeneous polynomial of degree n+2, (2) holds if the linear Hamiltonian vectorfield X_{f_0} is semisimple. The following theorem states that (2) also holds if X_{f_0} has only periodic integral curves.

Proposition 1.1. Let f be a smooth function on the symplectic manifold (M,Ω). If the Hamiltonian vectorfield X_f of f has only periodic integral curves, then

$$C^\infty(M) = \ker L_f \oplus \operatorname{im} L_f.$$

Proof. We study the equation

$$L_f h = g \tag{8}$$

where f, g, h $\in C^\infty(M)$. Let ϕ_t be the flow of X_f and suppose that $T = T(m) > 0$ is the period of the integral curve of X_f through m, that is, $\phi_{T(m)} = m$ for all $m \in M$. Then by a theorem of Gordon (1969), T is a smooth function of f. Thus $T : M \to \mathbb{R}$ is a smooth function. Since $\phi_t(m)$ and m lie on the same periodic orbit of X_f,

$$T(\phi_t(m)) = T(m) \text{ for all } m \in M. \tag{9}$$

It follows from the definition of Lie derivative that (8) is equivalent to the ordinary differential equation

$$\frac{d}{dt} h(\phi_t(m)) = g(\phi_t(m)). \tag{10}$$

Integrating (10) over a period gives

$$h(\phi_T(m)) - h(\phi_0(m)) = \int_0^T g(\phi_\tau(m))d\tau.$$

But $\phi_T(m) = \phi_0(m) = m$. So, if g is in the image of L_f, then the average $\overline{g}$ of g over the integral curves of X_f must vanish, that is,

$$\bar{g}(m) = \frac{1}{T}\int_0^T g(\phi_\tau(m))d\tau .$$

Next suppose that $\bar{g} = 0$ and let

$$h(m) = \frac{1}{T}\int_0^T \tau g(\phi_\tau(m))d\tau . \tag{11}$$

The following calculation shows that h defined by (11) solves (8). Since (9) holds, we have

$$h(\phi_t(m)) = \frac{1}{T}\int_0^T \tau g(\phi_{\tau+t}(m))d\tau$$

$$= \frac{1}{T}\int_t^{T+t} (u-t)g(\phi_u(m))du . \tag{12}$$

Differentiating (12) gives

$$\frac{d}{dt}h(\phi_t(m)) = \frac{1}{T}(T+t-t)g(\phi_{T+t}(m)) - \frac{1}{T}(t-t)g(\phi_t(m)) - \frac{1}{T}\int_t^{T+t} g(\phi_u(m))du$$

$$= g(\phi_t(m)) - \bar{g}(m), \text{ since } \phi_T(m) = m$$

$$= g(\phi_t(m)), \text{ since } \bar{g} = 0 \text{ by hypothesis.}$$

Therefore we have shown that

$$\text{im } L_f = \{g \in C^\infty(M) \mid \bar{g} = 0\}$$

which is a closed subspace of $C^\infty(M)$. Next we show that $g \in \ker L_f$. Since $T(\phi_t(m)) = T(m)$,

$$\bar{g}(\phi_t(m)) = \frac{1}{T}\int_t^{T+t} g(\phi_u(m))du$$

which differentiated gives

$$\frac{d}{dt}\bar{g}(\phi_t(m)) = \frac{1}{T}(g(\phi_t(\phi_T(m))) - g(\phi_t(m)))$$

$$= 0$$

because $\phi_T(m) = m$. Thus $\bar{g} \in \ker L_f$. To show the converse, suppose that $g \in \ker L_f$. Then $\frac{d}{dt} g(\phi_t(m)) = 0$; which implies $g(\phi_t(m)) = g(m)$. Consequently $\bar{g} = g$, that is, $\bar{g} \in \ker L_f$. Thus

$$\ker L_f = \{\bar{g} \in C^\infty(M) \mid g \in C^\infty(M)\}$$

which is a closed subspace of $C^\infty(M)$. Now $\ker L_f \cap \operatorname{im} L_f = \{0\}$. Since every smooth function g may be written as $g = \bar{g} + (g-\bar{g})$ where $\bar{g} \in \ker L_f$ and $(g-\bar{g}) \in \operatorname{im} L_f$ and the projection $\pi : C^\infty(M) \to \ker L_f : g \to \bar{g}$ is continuous, we have shown that $C^\infty(M)$ is the topological direct sum of $\ker L_f$ and $\operatorname{im} L_f$. □

2. NORMAL FORM: SPECIAL CASE

Here we determine the normal form of a smooth formal power series Hamiltonian $K_\varepsilon = \sum_{n\geqslant 0} \frac{\varepsilon^n}{n!} K_n$ where K_0 is the Hamiltonian on TS^n for geodesics on S^n.

We begin with some basic definitions. Let $<,>$ be the Euclidean inner product on $\mathbb{R}^{n+1}$ with corresponding norm $\| \ \|$. In (x,y) co-ordinates on $T\mathbb{R}^{n+1}$ the standard 1-form $\theta = <y,dx>$ and the standard 2-form $\omega = d\theta$. Let $S^n = \{x \in \mathbb{R}^{n+1} \mid \|x\|^2 = 1\}$ be the standard n-sphere and $T^+S^n = \{(x,y) \in T\mathbb{R}^{n+1} \mid \|x\|^2 = 1, <x,y> = 0, \|y\| > 0\}$ be the tangent bundle TS^n of S^n minus the zero section. Set $\Omega^+ = \omega | T^+S^n$. Then Ω^+ is a symplectic form on T^+S^n. On the symplectic manifold (T^+S^n, Ω^+) consider the Hamiltonian $K = K_0 : T^+S^n \to \mathbb{R}: (x,y) \to \|y\|$. Then K is smooth and the corresponding Hamiltonian vectorfield X_K

$$\frac{dx}{ds} = \frac{y}{\|y\|}$$
$$\frac{dy}{ds} = - \|y\| x \qquad (1)$$

is the geodesic vectorfield on T^+S^n. The flow ϕ^K of X_K is

$$\phi_s^K(x,y) = \begin{pmatrix} \cos(s)I_{n+1} & \frac{1}{\|y\|}\sin(s)I_{n+1} \\ -\|y\|\sin(s)I_{n+1} & \cos(s)I_{n+1} \end{pmatrix} \tag{2}$$

where I_{n+1} is the $n+1\times n+1$ identity matrix. The image of the integral curves of X_K under the bundle projection to S^n are geodesics on S^n.

Since $\phi_{2\pi}^K = id_{T^+S^n}$, all the integral curves of X_K are periodic. Thus by proposition 1.1, the Hamiltonian K_ε is in normal form with respect to $K_0 = K$ if and only if $L_K K_n = 0$ for all $n \geqslant 0$, that is, if and only if K_n are integrals of X_K. Hence to know what terms appear in the normal form of K we need to know all the smooth integrals of X_K.

We start by finding those integrals of X_K which arise from the $SO(n+1,\mathbb{R})$ symmetry of S^n. Consider the linear action $\sigma : SO(n+1,\mathbb{R})\times\mathbb{R}^{n+1} \to \mathbb{R}^{n+1} : A,x \to Ax$. Then σ lifts to a linear action $\hat{\sigma}$ of $SO(n+1,\mathbb{R})$ on $T\mathbb{R}^{n+1}$ defined by $\hat{\sigma} : SO(n+1,\mathbb{R})\times\mathbb{R} \to$ $\to T\mathbb{R}^{n+1} : A,(x,y) \to (Ax,Ay)$. $\hat{\sigma}$ is a symplectic action, that is, $\hat{\sigma}_A^*\omega = \omega$ for all $A \in SO(n+1,\mathbb{R})$, because $\hat{\sigma}_A^*\theta = \langle Ay,dAx\rangle =$ $= \langle Ay,Adx\rangle = \langle y,dx\rangle = \theta$. The $SO(n+1,\mathbb{R})$ momentum mapping of the $\hat{\sigma}$ action is $J : T\mathbb{R}^{n+1} \to so(n+1,\mathbb{R})^*$ where $J(x,y)a = \frac{1}{2}\langle y,ax\rangle$ and $a \in so(n+1,\mathbb{R})$. Since T^+S^n is invariant under $\hat{\sigma}$, $\hat{\sigma}$ restricts to a symplectic action $\hat{\Sigma}$ on (T^+S^n,Ω^+). The $SO(n+1,\mathbb{R})$ momentum mapping of the $\hat{\sigma}$ action is $j = J|T^+S^n$. Since K is invariant under $\hat{\sigma}$, j is an $so(n+1,\mathbb{R})^*$ valued integral of X_K. In particular, choosing the basis of $so(n+1,\mathbb{R})^*$ which is dual to the basis of $so(n+1,\mathbb{R})$ defined by

$$a_{ij} = \begin{pmatrix} \mathbf{0} & \overset{i}{0} & \overset{j}{1} & \mathbf{0} \\ \mathbf{0} & -1 & 0 & \mathbf{0} \end{pmatrix}\begin{matrix} i \\ j \end{matrix} \qquad 1 \leqslant i < j \leqslant n+1$$

gives

$$j(x,y) = (\lambda_{ij}) = (x_i y_j - x_j y_i)|T^+S^n \qquad 1 \leqslant i < j \leqslant n+1. \tag{3}$$

Note that λ_{ij} are integrals of X_K.

The problem remains: how do we show that the only integrals of X_K are smooth functions of the λ_{ij}? The solution is found by first converting the geodesic vectorfield on T^+S^n into the harmonic oscillator vectorfield on a certain cone C^+ in $T\mathbb{R}^{n+1}$, and then finding all the smooth integrals of the harmonic oscillator vectorfield. Let C^+ be the cone in $T\mathbb{R}^{n+1}$ defined by

$$\begin{cases} \|\xi\|^2 - \|\eta\|^2 = 0 \\ \langle\xi,\eta\rangle = 0 \qquad\qquad (\xi,\eta) \in T\mathbb{R}^{n+1}. \\ \|\eta\| > 0 \end{cases}$$

Put

$$\psi : T^+S^n \to C^+ : (x,y) \to (\xi,\eta) = (\|y\|x,y). \tag{4}$$

Then ψ is a smooth diffeomorphism of T^+S^n onto C^+. Define a vectorfield Y on C^+ by pushing X_K forward by ψ, that is, Y = $= \psi_* X_K = T\psi o X_K o \psi^{-1}$. Then a short calculation shows that Y is

$$\begin{aligned} \frac{d\xi}{ds} &= \eta \\ \frac{d\eta}{ds} &= -\xi \end{aligned} \qquad \text{where } (\xi,\eta) \in C^+. \tag{5}$$

Consider the harmonic oscillator Hamiltonian $G : T\mathbb{R}^{n+1} \to \mathbb{R}$: $(\xi,\eta) \to \frac{1}{2}(\|\xi\|^2 + \|\eta\|^2)$. Since C^+ is an invariant manifold of X_G, we see that Y is the restriction of X_G to C^+. Thus integrals of Y are just integrals of X_G restricted to C^+.

Next we prove the converse.

<u>Lemma</u> 2.1. Every smooth function on C^+ which is an integral of Y is the restriction to C^+ of a smooth integral of X_G on $T\mathbb{R}^{n+1}$.

<u>Proof</u>: Recall that a function on C^+ is smooth if it is the restriction to C^+ of a smooth function on $T\mathbb{R}^{n+1}$, that is, if $I = \{h \in C^\infty(T\mathbb{R}^{n+1}) \mid h|C^+ = 0\}$, then $C^\infty(C^+) = C^\infty(T\mathbb{R}^{n+1})/I$. Observe that I is a closed vector subspace of $C^\infty(T\mathbb{R}^{n+1})$. To

prove the lemma it suffices to show that if $f \in C^{\infty}(T\mathbb{R}^{n+1})$ and $h = L_G f \in I$, then $f \in \ker L_G + I$. Because I is closed and C^+ is invariant under the flow ϕ_t^G of X_G, using Riemann sums, it follows that if $h \in I$ then

$$H = \frac{1}{T}\int_0^T \tau(\phi^G)^* h d\tau \in I.$$

Using 1.11 we get

$$L_G(f - H) = L_G f - L_G H = h - h = 0,$$

that is,

$$f - H \in \ker L_G.$$

Hence $f \in \ker L_G + I$, since $H \in I$. □

Thus we need to find all smooth integrals of the harmonic oscillator. We start with

<u>Lemma</u> 2.2. All polynomials integrals of the harmonic oscillator vectorfield X_G on $T\mathbb{R}^{n+1}$ are polynomials in the $(n+1)^2$ quadratic integrals

$$\left\{ \begin{array}{ll} \xi_i\xi_j + \eta_i\eta_j & 1 \leqslant i \leqslant j \leqslant n+1 \\ \xi_i\eta_j - \xi_j\eta_i & 1 \leqslant i < j \leqslant n+1. \end{array} \right. \tag{6}$$

<u>Proof</u>: Introducing complex conjugate co-ordinates

$$\zeta = \xi + i\eta \qquad\qquad \overline{\zeta} = \xi - i\eta$$

the harmonic oscillator Hamiltonian G becomes $\tilde{G} = \frac{1}{2}\sum_{m=1}^{n+1} \zeta_m\overline{\zeta}_m$. Also the harmonic oscillator vectorfield X_G becomes the vectorfield $\tilde{X}_{\tilde{G}}$

$$\begin{aligned} \frac{d\zeta}{ds} &= 2i\,\frac{\partial\tilde{G}}{\partial\overline{\zeta}} = i\zeta \\ \frac{d\overline{\zeta}}{ds} &= -\,2i\,\frac{\partial\tilde{G}}{\partial\zeta} = -i\overline{\zeta}. \end{aligned} \tag{7}$$

Let $\zeta^r\bar{\zeta}^s$ be the monomial $\zeta_1^{r_1}\zeta_2^{r_2}\dots\zeta_{n+1}^{r_{n+1}}\bar{\zeta}_1^{s_1}\bar{\zeta}_2^{s_2}\dots\bar{\zeta}_{n+1}^{s_{n+1}}$ of degree $|r| + |s| = \sum_{m=1}^{n+1} (r_m+s_m)$. To find all polynomial integrals of X_G it suffices to determine all monomial integrals of $\tilde{X}_{\tilde{G}}$. The monomial $\zeta^r\bar{\zeta}^s$ is an integral of $X_{\tilde{G}}$ if and only if

$$0 = L_{\tilde{G}}\zeta^r\bar{\zeta}^s = i(\sum_{m=1}^{n+1} \zeta_m \frac{\partial}{\partial\zeta_m} - \bar{\zeta}_m \frac{\partial}{\partial\bar{\zeta}_m})\zeta^r\bar{\zeta}^s$$

that is, if and only if $|r| = |s|$. Suppose that $|r| + |s| = 2$ and $|r| = |s|$, then $|r| = |s| = 1$. Therefore exactly one component r_i or r and s_j of s is nonzero and equal to 1. Hence the quadratic monomial integrals of X_G are

$$\zeta_i\bar{\zeta}_j \qquad 1 \leqslant i \leqslant j \leqslant n + 1 \tag{8}$$

which in the original (ξ,η) co-ordinates are

$$\mathrm{Re}\ \zeta_i\bar{\zeta}_j = \xi_i\xi_j + \eta_i\eta_j \qquad 1 \leqslant i \leqslant j \leqslant n + 1$$

$$\mathrm{Im}\ \zeta_i\bar{\zeta}_j = \xi_i\eta_j - \xi_j\eta_i \qquad 1 \leqslant i < j \leqslant n + 1.$$

The following argument shows that the monomial integral $\zeta^r\bar{\zeta}^s$ with $|r| = |s|$ of X_G can be written as a product of quadratic monomial integrals. Write the components of r, s in nonincreasing order

$$r_{\pi(1)} \geqslant r_{\pi(2)} \geqslant \dots \geqslant r_{\pi(n+1)}$$

$$s_{\pi(1)} \geqslant s_{\pi(2)} \geqslant \dots \geqslant s_{\pi(n+1)}.$$

Let $a_i = r_{\pi(i)}$, $b_j = s_{\pi(j)}$ and let u, respectively v, be the largest index such that $a_u > 0$, respectively $b_v > 0$. Put $Z_i = \zeta_{\pi(i)}$ and $W_j = \bar{\zeta}_{\rho(j)}$. Then:

$$\zeta^r\bar{\zeta}^s = Z^aW^b = Z_1^{a_1}Z_2^{a_2}\dots Z_u^{a_u}W_1^{b_1}W_2^{b_2}\dots W_v^{b_v}$$

$$= \begin{cases} M(Z_uW_v)^{b_v}, & \text{if } a_u \geqslant b_v \\ N(Z_uW_v)^{a_u}, & \text{if } b_v \geqslant a_u \end{cases}$$

where $M = Z_1^{a_1} \ldots Z_u^{a_u - b_v} W_1^{b_1} \ldots W_{v-1}^{b_{v-1}}$ and $N =$
$= Z_1^{a_1} \ldots Z_{u-1}^{a_{u-1}} W_1^{b_1} \ldots W_c^{b_v - a_u}$.

Since

$$|r| = |a| = a_1 + \ldots + a_u = |s| = |b| = b_1 + \ldots + b_v,$$

we have

$$\begin{cases} |a| - b_v = |b| - b_v, \text{ if } a_u \geqslant b_v \\ |a| - a_u = |b| - a_u, \text{ if } b_v \geqslant a_u. \end{cases} \tag{9}$$

The above argument may be repeated on the monomials M and N because (9) implies that M and N are integrals of X_G. Since M and N are of degree lower than $|r| + |s|$, after a finite number of repetitions, $\zeta^r \bar{\zeta}^s$ with $|r| = |s|$ has been written as a product of quadratic integrals of $X_{\tilde{G}}$. □

The flow of the harmonic oscillator vectorfield X_G on $T\mathbb{R}^{n+1}$ defines an algebraic action Φ of $S^1 = SO(2,\mathbb{R})$ on $T\mathbb{R}^{n+1}$ given by $\Phi : SO(2,\mathbb{R}) \times T\mathbb{R}^{n+1} \to T\mathbb{R}^{n+1}$:

$$\begin{pmatrix} a & b \\ -b & a \end{pmatrix}, (\xi,\eta) \to \begin{pmatrix} aI_n & bI_n \\ -bI_n & aI_n \end{pmatrix} \begin{pmatrix} \xi \\ \eta \end{pmatrix}$$

where $a^2 + b^2 = 1$. The conclusion of lemma 2.2 can be rephrased as follows: the algebra of polynomials which are invariant under the action Φ is generated the $(n+1)^2$ quadratic polynomials given by (6). Because $SO(2,\mathbb{R})$ is a compact Lie group and the action is linear, we may apply a theorem of Schwarz (1975) to obtain

<u>Proposition</u> 2.2. Every C^∞ integral of the harmonic oscillator vectorfield X_G on $T\mathbb{R}^{n+1}$ is a C^∞ function of the quadratic integrals of X_G.

Pulling back the quadratic integrals of X_G restricted to the cone C^+ by the diffeomorphism ψ (see (4)) gives the integrals

$$\begin{cases} (\|y\|^2 x_i x_j + y_i y_j) | T^+S^n & 1 \leq i \leq j \leq n+1 \\ (\|y\| (x_i y_j - x_j y_i)) | T^+S^n & 1 \leq i \leq j \leq n+1 \end{cases} \tag{10}$$

of the geodesic vectorfield X_K on T^+S^n. Since

$$\|y\|^2 x_i x_j + y_i y_j = \sum_{\substack{k=1 \\ k \neq i,j}}^{n+1} (x_i y_k - x_k y_i)(x_j y_k - x_k y_j) +$$

$$+ (x_i y_j + x_j y_i) \sum_{k=1}^{n+1} x_k y_k + y_i y_j (1 - \sum_{k=1}^{n+1} x_k^2)$$

we have

$$(\|y\|^2 x_i x_j + y_i y_j) | T^+S^n = (\sum_{\substack{k=1 \\ k \neq i,j}}^{n+1} (x_i y_k - x_k y_i)(x_j y_k - x_k y_j)) | T^+S^n .$$

Moreover

$$\|y\| \, | T^+S^n = \Big(\sum_{1 \leq i < j \leq n+1} (x_i y_j - x_j y_i)^2 \Big)^{\frac{1}{2}} | T^+S^n .$$

Consequently the integrals of X_K given by (10) are smooth functions of the integrals

$$\lambda_{ij} = (x_i y_j - x_j y_i) | T^+S^n . \tag{11}$$

Therefore using proposition 2.2 we have proved

<u>Proposition</u> 2.3. Every C^∞ integral of the geodesic vectorfield X_K on T^+S^n is a C^∞ function of the integrals λ_{ij} $1 \leq i < j \leq n+1$.

An immediate consequence of propositions 2.3 and 1.1 is that the smooth Hamiltonian K_ε is in normal form with respect to $K_0 = K$ if and only if for every $n \geq 0$ K_n is a smooth function of λ_{ij}.

The case of proposition 2.3 when n = 3 is of special interest because of the relation between the bounded orbits of the Kepler problem on $T(\mathbb{R}^3 - \{0\})$ and geodesics on T^+S^3. To explain this relationship we follow Moser (1970) with some slight modifications. On $(T(\mathbb{R}^3-\{0\}), \sum_i dp_i \wedge dq_i)$ consider the Kepler Hamiltonian

$$M_0(q,p) = \tfrac{1}{2}\|p\|^2 - \frac{\mu}{\|q\|}, \ \mu > 0. \tag{12}$$

To convert (12) to the geodesic Hamiltonian K_0 on T^+S^3 we perform the following transformations. First, because we are only considering bounded orbits of M_0, add a positive constant $\tfrac{1}{2}h$ to M_0. Second, change the time scale from t to s according to

$$\frac{ds}{dt} = \frac{\sqrt{h}}{\|q\|} \tag{13}$$

and add a constant $\frac{\mu}{\sqrt{h}}$. The new time scale s is just the eccentric anomaly (Pollard 1966), and the new Hamiltonian is

$$\tilde{M}_0(q,p) = \frac{1}{\sqrt{h}}(\|q\| M_0(q,p) + \tfrac{1}{2}h) + \mu). \tag{14}$$

Third, interchange positions and momenta by

$$\begin{cases} \xi = -p \\ \eta = q. \end{cases}$$

The resulting Hamiltonian on $(T\mathbb{R}^3, \sum_i d\xi_i \wedge d\eta_i)$ is

$$\hat{M}_0(\xi,\eta) = \frac{1}{2\sqrt{h}}\|\eta\|(\|\xi\|^2 + h). \tag{15}$$

Fourth, apply a symplectic extension Ψ_h of stereographic projection from the north pole $p = (0,0,0,1)$ of $S^3 \subseteq \mathbb{R}^4$, where $\Psi_h : T(S^3 - \{p\}) \to T\mathbb{R}^3 : (x,y) \to (\xi,\eta)$ and

$$\xi_i = \sqrt{h}\,\frac{x_i}{(1 - x_4)} \qquad \eta_i = \frac{1}{\sqrt{h}}(y_i(1 - x_4) + x_i y_4). \tag{16}$$

The new Hamiltonian on $T(S^3 - \{p\})$ is

$$K_0(x,y) = \hat{M}_0 \circ \Psi_h(x,y) = \|y\|. \tag{17}$$

Because the transformations which convert M_0 to K_0 map the level set $M_0^{-1}(L)$ diffeomorphically onto $K_0^{-1}(L)$ (where $L = \frac{\mu}{\sqrt{h}}$) and also take bounded linear orbits (which reach 0 in finite time) to geodesics on S^3 which pass through the north pole p, the domain T^+S^3 corresponds precisely to those initial conditions in $T(\mathbb{R}^3 - \{0\})$ which give rise to bounded orbits of the Kepler vectorfield. Applying the inverse transformations to the integrals $\lambda_{ij}|K_0^{-1}(L)$ shows that on $K_0^{-1}(L)$

$$\begin{pmatrix} \lambda_{41} \\ \lambda_{42} \\ \lambda_{43} \end{pmatrix} = \begin{pmatrix} x_4y_1 - x_1y_4 \\ x_4y_2 - x_2y_4 \\ x_4y_3 - x_3y_4 \end{pmatrix}$$

corresponds to

$$\frac{1}{\sqrt{h}}(p\times(q\times p) - \frac{\mu q}{\|q\|}) = LA \tag{18}$$

where A is the eccentric axis vector which points along the axis of the ellipse toward the perigee and has length equal to the eccentricity of the ellipse; moreover, on $K_0^{-1}(L)$

$$\begin{pmatrix} \lambda_{23} \\ -\lambda_{13} \\ \lambda_{12} \end{pmatrix} = \begin{pmatrix} x_2y_3 - x_3y_2 \\ x_3y_1 - x_1y_3 \\ x_1y_2 - x_2y_1 \end{pmatrix}$$

corresponds to

$$q\times p = J \tag{19}$$

where J is the angular momentum. Changing the level of K_0 from L to $\rho^{-1}L$ multiplies the right hand side of (18) and (19) by ρ. From proposition 2.3 we obtain

<u>Corollary</u> 2.4. Every C^∞ integral on the bounded Keplerian orbits in $\mathbb{R}^3$ is a C^∞ function of the energy and the components of the eccentric axis and the angular momentum vectors.

3. REDUCTION AND INVARIANT THEORY

In this section we apply invariant theory to carry out the reduction process on a smooth formal power series Hamiltonian $K_\varepsilon : T^+S^3 \to \mathbb{R}$ which has the following properties:

1) K_ε is in normal form with respect to the geodesic Hamiltonian $K_0 = K : T^+S^3 \to \mathbb{R} : (x,y) \to \|y\|$; and
2) K_ε is invariant under the axial symmetry Φ defined by $\Phi : S^1 \times T^+S^3 \to T^+S^3 : t,(x,y) \to (R_t x, R_t y)$ where $R_t =$

$$\begin{pmatrix} \cos t & \sin t & & \\ -\sin t & \cos t & & 0 \\ & & 1 & 0 \\ 0 & & 0 & 1 \end{pmatrix}.$$

Note that under symplectic stereographic projection Ψ_h (see 2.16) $\Psi_h^{-1} \circ \Phi_t \circ \Psi_h$ is a counterclockwise rotation about the z-axis in $\mathbb{R}^3$ through an angle t.

We begin by performing reduction on the S^1 action defined by the geodesic flow ϕ^K on (T^+S^3, Ω^+). Because K_ε is in normal form with respect to $K = K_0$, K_0 is an integral of X_{K_ε} which implies that the level set $K_0^{-1}(L)$ is invariant under the flow of X_{K_ε}. Since $\phi^K_{2\pi} = \mathrm{id}$, ϕ^K defines a fixed point free, proper, S^1 action on $K_0^{-1}(L)$. Hence the S^1 orbit space $P_L = K_0^{-1}(L)/S^1$ exists and is a smooth symplectic manifold (Marsden & Weinstein 1974). We now explicitly construct the reduced phase space P_L, its symplectic form, and the reduced Hamiltonian vectorfield.

From proposition 2.3 we know that the integrals λ_{ij} $1 \leqslant i < j \leqslant 4$ generate the algebra of smooth functions which are invariant under ϕ^K. A short calculation shows that λ_{ij} satisfy the following relations:

$$\begin{cases} 0 = \lambda_{41}\lambda_{23} - \lambda_{13}\lambda_{42} + \lambda_{12}\lambda_{43} \\ L^2 = \lambda_{41}^2 + \lambda_{42}^2 + \lambda_{43}^2 + \lambda_{23}^2 + \lambda_{13}^2 + \lambda_{12}^2 \end{cases} \tag{1}$$

which define the reduced space P_L. Introducing the reduced variables

$$\begin{aligned} z_1 &= \lambda_{23} + \lambda_{41} & \zeta_1 &= \lambda_{23} - \lambda_{41} \\ z_2 &= -\lambda_{12} + \lambda_{43} & \zeta_2 &= -\lambda_{12} + \lambda_{43} \\ z_3 &= -\lambda_{13} + \lambda_{42} & \zeta_3 &= -\lambda_{13} - \lambda_{42} \end{aligned}$$

we see that (1) is equivalent to

$$\begin{cases} z_1^2 + z_2^2 + z_3^2 = L^2 \\ \zeta_1^2 + \zeta_2^2 + \zeta_3^2 = L^2, \end{cases} \tag{2}$$

which shows that P_L is $S_L^2 \times S_L^2$. To find the symplectic form Ω_L on P_L, we first find the reduced Hamiltonian vectorfield X_{K_L} of the reduced Hamiltonian function K_L. Since K_ε is in normal form, by proposition 2.3 K_ε is a smooth function of λ_{ij} which in turn is a smooth function K_L of z,ζ. Note that

$$\begin{aligned} \{z_1,z_2\} &= z_3 & \{z_i,\zeta_j\} &= 0 & \{\zeta_1,\zeta_2\} &= \zeta_3 \\ \{z_2,z_3\} &= z_1 & & & \{\zeta_2,\zeta_3\} &= \zeta_1 \\ \{z_3,z_1\} &= z_2 & & & \{\zeta_3,\zeta_1\} &= \zeta_2 \end{aligned} \tag{3}$$

where $\{,\}$ is Poisson bracket on (T^+S^3,Ω_L^+). Since $K_L = K_L(z,\zeta)$, $X_{K_L} = (\{K_L,z\}, \{K_K,\zeta\})$ where

$$\begin{aligned} \{K_L,z_i\} &= \sum_{j=1}^{3} \left(\frac{\partial K_L}{\partial z_j}\{z_j,z_i\} + \frac{\partial K_L}{\partial \zeta_j}\{\zeta_j,z_i\}\right) \\ &= \sum_{j=1}^{3} \frac{\partial K_L}{\partial z_j}\{z_j,z_i\} \quad \text{using (3)} \end{aligned}$$

and similarly

$$\{K_L,\zeta_i\} = \sum_{j=1}^{3} \frac{\partial K_L}{\partial \zeta_j}\{\zeta_j,\zeta_i\}.$$

In terms of the vector product $\times$ on $\mathbb{R}^3$, X_{K_L} may be written as

$$\begin{cases} \dfrac{dz}{ds} = -z \times \mathrm{grad}_z K_L \\ \dfrac{d\zeta}{ds} = -\zeta \times \mathrm{grad}_\zeta K_L \end{cases} \quad \text{on } \mathbb{R}^6 \tag{4}$$

restricted to P_L (which is an invariant manifold of (4)). If we define a symplectic form Ω_L on P_L by

$$\Omega_L(z,\zeta)((z\times a,\zeta\times b),\ (z\times a',\zeta\times b')) = \langle z,a\times b\rangle + \langle \zeta,a'\times b'\rangle \tag{5}$$

where a, a', b, b' $\in \mathbb{R}^3$, then on (P_L,Ω_L) X_{K_L} is given by (5). Thus Ω_L is the induced symplectic form on the reduced phase P_L.

We now carry out the reduction process on (P_L,Ω_L) for the axial symmetry. A short calculation shows that the S^1 action $\ddot{\Phi}$ on (T^+S^3,Ω^+) induces an S^1 action $\check{\Phi}$ on $S^2_L\times S^2_L$ defined by $\check{\Phi} : S^1\times(S^2_L\times S^2_L) \to S^2_L\times S^2_L : t,(z,\zeta) \to (S_t z, S_t\zeta)$ where $S_t =$

$$\begin{pmatrix} \cos t & \sin t & 0 \\ -\sin t & \cos t & 0 \\ 0 & 0 & 1 \end{pmatrix}.$$

Since $S_t \in SO(3,\mathbb{R})$, $\check{\Phi}$ is a symplectic S^1 action on (P_L,Ω_L) (see 2.5). In fact $\check{\Phi}$ is a Hamiltonian action with momentum mapping $F : S^2_L\times S^2_L \to \mathbb{R} : (z,\zeta) \to z_3 + \zeta_3$ as the following calculation shows:

$$\begin{aligned} X_F(z,\zeta) &= -(z\times\mathrm{grad}_z F,\zeta\times\mathrm{grad}_\zeta F) \\ &= (z_2,-z_1,0,\zeta_2,-\zeta_1,0) \\ &= \frac{d}{dt}\Big|_{t=0} \check{\Phi}_t(z,\zeta). \end{aligned}$$

Eventhough the S^1 action $\check{\Phi}$ has fixed points on $F^{-1}(-2H)$, all the orbits of $\check{\Phi}$ are closed. Hence the orbit space $P_{H,L} = F^{-1}(-2H)/S^1$ exists as a topological manifold. To construct $P_{H,L}$ we note that, after a short calculation, the algebra of polynomials invariant under $\check{\Phi}$ is generated by

$$\pi_1 = z_3 \qquad \pi_2 = \zeta_3 \qquad \pi_3 = z_1^2 + z_2^2$$
$$\pi_4 = \zeta_1^2 + \zeta_2^2 \qquad \pi_5 = z_1\zeta_2 - \zeta_1 z_2 \qquad \pi_6 = z_1\zeta_1+z_2\zeta_2 \tag{6}$$

which satisfies the relations

$$\pi_1^2 + \pi_3 = L^2 \qquad \pi_2^2 + \pi_4 = L^2 \qquad \pi_5^2+\pi_6^2 = \pi_3\pi_4. \tag{7}$$

Introducing the reduced variables

$$\sigma_1 = \tfrac{1}{2}(\pi_1-\pi_2) \qquad \sigma_2 = \pi_5 \qquad \sigma_3 = \pi_6, \tag{8}$$

we see that (7) is equivalent to

$$B(\sigma) = \tfrac{1}{2}(\sigma_2^2 + \sigma_3^2) - \tfrac{1}{2}(L^2 - (\sigma_1-H)^2)(L^2 - (\sigma_1+H)^2) \tag{9}$$

which defines the reduced space $P_{H,L}$ as an algebraic subvariety of $\mathbb{R}^3$. From (9) it follows that $P_{H,L}$ is diffeomorphic to S^2 if $0 < |H| < L$ and is homeomorphic to a point $(0,0,0)$ if $|H| = L$ or a double cone over S^1 with vertices at $\pm(0,0,L)$ if $H = 0$.

From now on we will assume that $0 < |H| < L$ so that $P_{H,L}$ is a smooth manifold. Since $K_\varepsilon(x,y) = K_L(z,\zeta)$ is invariant under Φ and $K_\varepsilon \circ \Phi_t = K_L \circ \check{\Phi}_t$, K_L is invariant under $\check{\Phi}$. Hence K_L induces a function $K_{H,L}$ on $P_{H,L}$. To find $X_{K_{H,L}}$ on $P_{H,L}$ note that

$$\begin{aligned} \{\{\sigma_2,\sigma_2\}\} &= -\frac{\partial B}{\partial\sigma_3} \\ \{\{\sigma_2,\sigma_3\}\} &= -\frac{\partial B}{\partial\sigma_1} \\ \{\{\sigma_3,\sigma_1\}\} &= -\frac{\partial B}{\partial\sigma_2} \end{aligned} \tag{10}$$

where $\{\{,\}\}$ is Poisson bracket on (P_L,Ω_L). Since $K_{H,L} =$ $= K_{H,L}(\sigma)$, $X_{K_{H,L}} = \{\{K_{H,L},\sigma\}\}$ where

$$\{\{K_{H,L},\sigma_i\}\} = \sum_{j=1}^{3} \frac{\partial K_{H,L}}{\partial\sigma_j} \{\{\sigma_j,\sigma_i\}\}. \tag{11}$$

Using the vector product $\times$ on $\mathbb{R}^3$, $X_{K_{H,L}}$ may be written as

$$\frac{d\sigma}{ds} = \text{grad } B \times \text{grad } K_{H,L} \text{ on } \mathbb{R}^3 \tag{12}$$

restricted to $P_{H,L}$ (which is an invariant manifold of (12)). If we define the symplectic form $\Omega_{H,L}$ on $P_{H,L}$ by

$$\Omega_{H,L}(\sigma)(a,b) = -(\langle \text{grad } B, \text{grad } B \rangle)^{-1} \langle \text{grad } B, a \times b \rangle \tag{13}$$

where a, b $\in T_\sigma P_{H,L} = \ker dB(\sigma)$, then on $(P_{H,L}, \Omega_{H,L})$, $X_{K_{H,L}}$ is given by (12). Hence $\Omega_{H,L}$ is the induced symplectic form on $P_{H,L}$. This completes the reduction of K_ε to a one degree of freedom Hamiltonian $K_{H,L}$ on the 2-sphere $(P_{H,L}, \Omega_{H,L})$ when $0 < |H| < L$.

References

Abraham, R. & Marsden, J., Foundations of Mechanics, 2nd ed., Benjamin/Cummings, Reading, Mass., 1978.

Cushman, R. 1982, Reduction, Brouwer's Hamiltonian, and the critical inclination, preprint No. 243, University of Utrecht.

Deprit, A. 1969, Celest. Mech., 1, 1.

Deprit, A. 1982, Celest. Mech., 26, 9.

Gordon, W.B. 1969, J. Math. Mech., 19, 111.

Marsden, J. & Weinstein, A. 1974, Rep. Math. Phys., 5, 121.

Moser, J. 1970, Comm. Pure Appl. Math., 23, 609.

Pollard, H., Mathematical Introduction to Celestial Mechanics Prentice Hall, Englewood Cliffs, N.J., 1966.

Schwarz, G. 1975, Topology, 14, 63.

MAGNETIC SOLUTION OF YANG-MILLS EQUATIONS AND THE MOTION OF CLASSICAL PARTICLE

M. Carmeli and Kh. Huleihil
Center for Theoretical Physics
Ben-Gurion University of the Negev
Beer-Sheva 84105, Israel

A magnetic solution of the Yang-Mills equations is presented and examined by investigating the equations of motion of a classical particle moving in the field. The motion is considered in both the Minkowskian and the internal spaces. The solution is of type D_p in Carmeli's classification, class A in Castillejo-Kugler-Roskies's classification scheme, and case 3_b in that of Wang-Yang's.

1. INTRODUCTION

In this paper we present a magnetic solution to the classical Yang-Mills field equations, and find the equations of motion of a classical particle moving in this field. To get the magnetic solution we consider the fields of class A in Castillejo-Kugler-Roskies's classification (1), and we transform these fields to the null-tetrad version of Carmeli-Charach-Kaye (2).

Consequently, a simplified version of the Yang-Mills fields and equations is derived. The obtained simplified equations then yield a special condition for the unknown functions. This condition subsequently leads to a solution which belongs to class D_p in Carmeli's classification (3,4), and to case 3_b in Wang-Yang's classification scheme (5).

The null-tetrad theory of the Yang-Mills equations was given by Carmeli, Charach, and Kaye (2), and also by Newman (6,7). The Yang-Mills dynamical variables, in the presence of gravitation, were presented in terms of new variables defined by means of a null-tetrad, using the spin coefficient method. The resulting equations then resemble the Newman-Penrose (8,9) version of Maxwell's equations extended to non-Abelian gauge groups.

S. Sternberg (ed.), Differential Geometric Methods in Mathematical Physics, 145–159.

The sourceless field equations for the SU(2) gauge theory (10), in the particular case of fields which belong to class A, are then given by

$$\left(\frac{\partial}{\partial u} - \frac{1}{2}\frac{\partial}{\partial r}\right)\beta_j = g\varepsilon_{kmj}\beta_k b_{11'm} \ , \tag{1.1a}$$

$$\frac{\partial}{\partial r}\beta_j = g\varepsilon_{kmj}\beta_k b_{00'm} \ , \tag{1.1b}$$

$$\frac{1}{\sqrt{2}r}\mathcal{D}\beta_j = g\varepsilon_{kmj}\beta_k b_{01'm} \ , \tag{1.1c}$$

$$\frac{1}{\sqrt{2}r}\bar{\mathcal{D}}\beta_j = g\varepsilon_{kmj}\beta_k b_{10'm} \ . \tag{1.1d}$$

Here k, m, j are isospin indices taking the values 1, 2, 3, and $b_{cd'k}$ are the potential isotriplet, related to the ordinary potential $b_{\mu k}$ by

$$b_{cd'k} = \sigma^{\mu}_{cd'} b_{\mu k} \ , \tag{1.2}$$

where $\sigma^{\mu}_{cd'}$ is a tetrad of null vectors. The indices c, d' take the values 0, 1 and 0', 1', respectively. Greek indices take the values 0, 1, 2, 3, a bar indicates complex conjugation, and our metric is (+---). In Eqs. (1.1) β_k is an isospin vector given by

$$\beta_k = (a+ib,\ ic,\ 0) \ , \tag{1.3}$$

where a, b, and c are real functions of the coordinates and are related to the field strengths $f^k_{\mu\nu}$ by

$$a(x) = f^1_{10} = - f^1_{01} \ , \tag{1.4a}$$

$$b(x) = f^1_{23} = - f^1_{32} \ , \tag{1.4b}$$

$$c(x) = f^2_{23} = - f^2_{32} \ . \tag{1.4c}$$

All the other components of the field strengths vanish. The operator $\mathcal{D}$ is a differential operator given by

$$\mathcal{D} = \frac{\partial}{\partial\theta} + \frac{i}{\sin\theta}\frac{\partial}{\partial\varphi} \ ,$$

the coordinate u = t-r is a retarded time coordinate, and g is the coupling constant.

From Eqs. (1.1) and the reality of the functions a, b, c, $b_{00'k}$ and $b_{11'k}$ it then follows that either a or $b_{00'k}$ and $b_{11'k}$ vanish. In the following we assume that a=0. Equations (1.1) then yield

$$b_{00'3} = \frac{1}{gc}\frac{\partial b}{\partial r} , \tag{1.5a}$$

$$b_{01'3} = \frac{1}{\sqrt{2}grc}\mathcal{D}b , \tag{1.5b}$$

$$b_{11'3} = \frac{1}{gc}\left(\frac{\partial}{\partial u} - \frac{1}{2}\frac{\partial}{\partial r}\right)b , \tag{1.5c}$$

$$b_{ab'1} = \frac{b}{c} b_{ab'2} , \tag{1.5d}$$

along with the condition that b^2+c^2 is a constant.

It is worth mentioning that using the null-tetrad method in gravitation brought a remarkable simplification into the Einstein field equations thus enabling Kinnersley to find all type D vacuum solutions (11). The extension to non-Abelian gauge theory of the spin coefficient method has also brought a significant simplification which enabled obtaining a monopole solution that includes both electric and magnetic charges (12), and to other new solutions (13), although new solutions may also be obtained through other methods utilizing classification schemes too (14).

The solution is discussed by studying the motion of a classical particle moving in the field. To this end we use the equations of motion which were derived by Wong (15) from quantum fields, taking appropriate limits, and independently by Drechsler and Rosenblum (16), who obtained the same equations of motion from energy and momentum conservation, and covariant charge conservation, following a procedure first introduced by Mathisson (17,18).

The equations of motion of the particle (in the Minkowskian space) in the Yang-Mills field are analogous to those of a charge in an ordinary electromagnetic field, and in the special case of our solution, the motion is analogous to that of a charge in an ordinary magnetic field which is directed along the x axis (19).

In Section 2 the solution of the field equations is presented. The equations of motion of a classical particle moving in the Yang-Mills field are subsequently given in Section 3. The solutions of these equations of motion in the particular Yang-Mills field, which is presented in Section 2, are given in Section 4.

2. SOLUTION OF THE FIELD EQUATIONS

A particular solution of the above set of equations and conditions is easily found to be given by

$$b(x) = K \cos \omega(\theta, \varphi) , \tag{2.1}$$

$$c(x) = K \sin \omega(\theta, \varphi) \quad , \tag{2.2}$$

where K is a constant, and ω is a real function of θ and φ. Then the field strengths

$$\chi_{abk} = \frac{1}{2} \varepsilon^{c'd'} \sigma^{\mu}_{ac'} \sigma^{\nu}_{bd'} f_{\mu\nu k}$$

can be presented in the Abelian form

$$\chi_{abk} = \frac{1}{g} \chi_{ab} \alpha_k \quad , \tag{2.3}$$

where χ_{ab} is given by

$$\chi_{00} = - \frac{i}{\sqrt{2}} (\cos\theta \cos\varphi - i \sin\varphi) \quad , \tag{2.4a}$$

$$\chi_{01} = \chi_{10} = \frac{i}{2} \sin\theta \cos\varphi \quad , \tag{2.4b}$$

$$\chi_{11} = \frac{1}{2} \bar{\chi}_{00} \quad , \tag{2.4c}$$

and α_k is an isospin vector,

$$\alpha_k = K[\cos \omega(\theta, \varphi), \sin \omega(\theta, \varphi), 0] \ . \tag{2.5}$$

From Eqs. (1.5), (2.1) and (2.2) we then obtain

$$b_{00'3} = b_{11'3} = 0 \quad , \tag{2.6a}$$

$$b_{01'3} = - \frac{1}{\sqrt{2}gr} \mathcal{D}\omega(\theta, \varphi) \ . \tag{2.6b}$$

It is easy to see that a general solution is given by

$$b(x) = K \cos \omega(u, r, \theta, \varphi) \quad ,$$

$$c(x) = K \sin \omega(u, r, \theta, \varphi) \quad ,$$

and thus

$$\alpha_k = K[\cos \omega(u, r, \theta, \varphi), \sin \omega(u, r, \theta, \varphi), 0] \ .$$

The choice of the particular solution given by Eqs. (2.1) and (2.2) is made in order to simplify the calculation of the potentials.

Now from Eqs. (2.3) and (2.4) one finds that

$$\bar{\chi}_{00k} - 2\chi_{11k} = 0 \ . \tag{2.7}$$

Then from the relations between χ_{abk} and $b_{ab'k}$ we get the following equation:

$$\frac{1}{\sqrt{2}r}\bar{\mathcal{D}}\left(b_{00'k} + 2b_{11'k}\right) - 2\frac{\partial}{\partial u} b_{10'k}$$

$$+ g\varepsilon_{mjk}\left(b_{00'j} + 2b_{11'j}\right)b_{10'm} = 0 \ . \qquad (2.8)$$

If we now assume that $b_{01'k}$ does not depend on u, and using Eqs. (2.6), we then obtain

$$b_{00'1} + 2b_{11'1} = \frac{1}{g}\, e^{f(u,r)}\cos\,\omega(\theta,\,\varphi)\ , \qquad (2.9a)$$

$$b_{00'2} + 2b_{11'2} = \frac{1}{g}\, e^{f(u,r)}\sin\,\omega(\theta,\,\varphi)\ , \qquad (2.9b)$$

where f(u,r) is an arbitrary real function of u and r. Taking the sum of χ_{00k} and $2\chi_{11k}$, one finds

$$\sqrt{2}\,\frac{i}{g}(\cos\theta\,\cos\varphi + i\,\sin\varphi)\alpha_k = \frac{1}{\sqrt{2}r}\bar{\mathcal{D}}\left(b_{00'k} - 2b_{11'k}\right)$$

$$- 2\left(\frac{\partial}{\partial r} + \frac{1}{r}\right)b_{10'k} + g\varepsilon_{mjk}b_{10'm}\left(b_{00'j} - 2b_{11'j}\right) \ . \qquad (2.10)$$

Under the assumption that $b_{00'k} = 2b_{11'k}$, the latter equation then yields

$$\left(\frac{\partial}{\partial r} + \frac{1}{r}\right)b_{01'k} = \frac{i}{\sqrt{2}g}\,(\cos\theta\,\cos\varphi - i\,\sin\varphi)\alpha_k\ , \qquad (2.11)$$

whose solution is given by

$$b_{01'k} = \frac{ir}{2\sqrt{2}g}\,(\cos\theta\,\cos\varphi - i\,\sin\varphi)\alpha_k + \frac{1}{\sqrt{2}gr}\,h_k(\theta,\,\varphi)\ , \qquad (2.12)$$

where h_k are complex functions of θ and φ satisfying

$$h_1(\theta,\,\varphi) = \cot\,\omega(\theta,\,\varphi)\ h_2(\theta,\,\varphi)\ , \qquad (2.13a)$$

$$h_3(\theta,\,\varphi) = -\,\mathcal{D}\omega\ . \qquad (2.13b)$$

From the real part of χ_{01k} one obtains the following condition on the function f(u,r):

$$\left(\frac{\partial}{\partial u} - \frac{\partial}{\partial r}\right)f(u,r) = 0\ , \qquad (2.14)$$

from which one finds that f(u,r) = f(u+r). From the imaginary part of χ_{01k} one obtains

$$\mathrm{Im}[\mathcal{D} + \cot\theta - \mathcal{D}(\ln\,\cos\omega)]\ \bar{h}_1(\theta,\,\varphi) = 0\ . \qquad (2.15)$$

Summarizing the above results, we get for the potentials

$$b_{00'k} = \frac{1}{g} e^{f(u+r)} \alpha_k , \tag{2.16a}$$

$$b_{01'k} = \frac{ir}{2\sqrt{2}g} (\cos\theta \cos\varphi - i \sin\varphi)\alpha_k + \frac{1}{\sqrt{2}gr} h_k(\theta, \varphi) , \tag{2.16b}$$

$$b_{11'k} = \frac{1}{2g} e^{f(u+r)} \alpha_k , \tag{2.16c}$$

and $b_{10'k}$ is the complex conjugate of $b_{01'k}$.

The above potentials, along with the field strengths given by Eqs. (2.3) and (2.4), constitute a solution of the Yang-Mills equations.

Using Minkowskian coordinates, the above solution can then be written in the form

$$b^k_\mu = \frac{1}{g}\left(A_\mu \alpha^k + B_\mu h^k_{Re} + D_\mu h^k_{Im}\right) , \tag{2.17}$$

$$f^k_{\mu\nu} = \frac{1}{g} f_{\mu\nu}\alpha^k , \tag{2.18}$$

where

$$A_\mu = \left(e^{f(t)}, 0, \tfrac{1}{2}z, -\tfrac{1}{2}y\right) , \tag{2.19}$$

$$B_\mu = \frac{1}{r^2 R}\left(0, xz, yz, -R^2\right) , \tag{2.20}$$

$$D_\mu = \frac{1}{rR}(0, -y, x, 0) , \tag{2.21}$$

$$f_{23} = - f_{32} = 1 \text{ (other components vanish)}. \tag{2.22}$$

Here $R^2 = x^2+y^2$, $r^2 = R^2+z^2$, h^k_{Re} and h^k_{Im} are the real and the imaginary parts of $h_k(\theta, \varphi)$, respectively.

The above solution describes a unidirectional constant field in space-time whose direction in the isospin space is determined by the vector α_k given by Eq. (2.5). Hence one should expect that the energy density of the field to be a constant throughout the space-time. This can easily be seen if one calculates the energy density

$$T_{00} = \frac{1}{2\pi} \sigma_0^{ac'} \sigma_0^{bd'} \chi_{abk}\bar{\chi}_{c'd'k} , \tag{2.23}$$

where $T_{\mu\nu}$ is the energy-momentum tensor. One then finds

$$T_{00} = \frac{K^2}{8\pi g^2} . \tag{2.24}$$

3. MOTION OF A CLASSICAL PARTICLE

More insight on the above solution can be obtained by investigating the equations of motion of a classical particle moving in the field. To this end one can use the equations of motion, recently developed by Wong (15) from quantum fields taking appropriate limits, and by Drechsler and Rosenblum (16) who derived the same equations of motion from energy and momentum conservation, and covariant charge conservation, following a procedure first given by Mathisson (17,18). We now summarize Wong's method.

Starting with the Lagrangian density of an interacting system of a Yang-Mills field and a particle with spin ½,

$$L = -\frac{1}{4}\,\underset{\sim}{f}^{\mu\nu}\cdot\underset{\sim}{f}_{\mu\nu} - \bar{\psi}\gamma^{\mu}(\partial_{\mu}-ig\underset{\sim}{b}_{\mu}\cdot\underset{\sim}{X})\psi - \frac{mc}{\hbar}\,\bar{\psi}\psi\ , \tag{3.1}$$

where ψ is the field of the particle, X_k are the generators of SU(2) satisfying the commutation relations

$$[X_i,\ X_j] = i\ \varepsilon_{ijk}X_k\ , \tag{3.2}$$

and γ_μ are Hermitian matrices satisfying the anticommutation relations

$$\{\gamma_\mu,\ \gamma_\nu\} = 2\delta_{\mu\nu}\ .$$

The field equations are then given by

$$\partial^{\nu}\underset{\sim}{f}_{\mu\nu} + g\underset{\sim}{b}^{\nu}\times\underset{\sim}{f}_{\mu\nu} = -\ ig\ \bar{\psi}\gamma_{\mu}\underset{\sim}{X}\psi\ , \tag{3.3}$$

$$\gamma^{\mu}\left(\partial_{\mu}-ig\underset{\sim}{X}\circ\underset{\sim}{b}_{\mu}\right)\psi + \frac{mc}{\hbar}\psi = 0\ , \tag{3.4}$$

where

$$\underset{\sim}{f}_{\mu\nu} = \partial_{\nu}\underset{\sim}{b}_{\mu} - \partial_{\mu}\underset{\sim}{b}_{\nu} - g\underset{\sim}{b}_{\mu}\times\underset{\sim}{b}_{\nu}\ . \tag{3.5}$$

(Notice that our $\underset{\sim}{f}$ has an opposite sign of Wong's.)

Regarding Eq. (3.4) as a one-particle Dirac equation for a particle with an isotopic spin in the given external c-number Yang-Mills field, it can be written in the form

$$i\hbar\,\frac{\partial\psi}{\partial t} = H\psi\ , \tag{3.6}$$

with

$$H = c\alpha_i(p_i-g\underset{\sim}{b}_i\cdot\underset{\sim}{I}) - imc^2\beta - gc\underset{\sim}{b}_0\cdot\underset{\sim}{I}\ , \tag{3.7}$$

where α_i, β are the Dirac matrices, $p_i = (\hbar/i)\partial_i$, and $\underset{\sim}{I} = \hbar\underset{\sim}{X}$ satisfying the commutation relations

$$[I_i, I_j] = i\hbar\, \varepsilon_{ijk} I_k \; . \tag{3.8}$$

In the Heisenberg picture, the following equations of motion are obtained:

$$\frac{dx_i}{dt} = \frac{i}{\hbar}\,[H, x_i] = c\alpha_i \; , \tag{3.9}$$

$$\frac{dp_i}{dt} = \frac{i}{\hbar}\,[H, p_i] = gc(\alpha_j \partial_i \underset{\sim}{b}_j + \partial_i \underset{\sim}{b}_0)\cdot \underset{\sim}{I} \; , \tag{3.10}$$

$$\frac{d\underset{\sim}{I}}{dt} = \frac{i}{\hbar}\,[H, \underset{\sim}{I}] = -\, g\left(\frac{dx_i}{dt}\,\underset{\sim}{b}_i + c\underset{\sim}{b}_0\right) \times \underset{\sim}{I} \; . \tag{3.11}$$

Define now the mechanical momenta by

$$\pi_i = p_i - g\underset{\sim}{b}_i \cdot \underset{\sim}{I} \; , \tag{3.12}$$

one then finds

$$\frac{d\pi_i}{dt} = g\left(\frac{dx_j}{dt}\,\underset{\sim}{f}_{ij} + c\underset{\sim}{f}_{i0}\right)\cdot \underset{\sim}{I} \; . \tag{3.13}$$

Comparing Eq. (3.13) with the corresponding equation for a charged particle moving in an external electromagnetic field, the following equation governing the world line $\xi^\mu(\tau)$ of the particle in space-time suggests itself:

$$m\ddot{\xi}^\mu = g\underset{\sim}{f}^{\mu\nu}\cdot \underset{\sim}{I}\dot{\xi}_\nu \; , \tag{3.14}$$

where the dot denotes differentiation with respect to the proper time τ. The right-hand side of Eq. (3.14) obviously represents a generalization of the Lorentz force. In addition, one obtains

$$\dot{\underset{\sim}{I}} + g\underset{\sim}{b}^\mu \times \underset{\sim}{I}\dot{\xi}_\mu = 0 \; . \tag{3.15}$$

Thus a particle is described by an internal vector $\underset{\sim}{I}$ as well as its space-time coordinates ξ^μ.

The isotopic spin current carried by a point particle, analogous to the electric current, is given by

$$g \int \underset{\sim}{I}(\tau)\dot{\xi}_\mu \delta^{(4)}[x - \xi(\tau)]\,d\tau \; . \tag{3.16}$$

Hence the field equations will be of the Yang-Mills form with the current $\underset{\sim}{j}_\mu$ being a sum of terms of the form (3.16) for each particle. Equations (3.14) and (3.15), along with the field equations, describe the interaction of a system of particles having isotopic spins.

An immediate consequence of Eq. (3.15) is

$$\frac{d}{d\tau} \underset{\sim}{I}^2 = 0 \ . \tag{3.17}$$

Hence the isotopic spin of each particle performs a precessional motion.

Now we give another approach where the equations of motion of a non-Abelian charge follow from the principle of energy and momentum, and the covariant current, conservation. The equations obtained are identical in structure to those derived above in the semiclassical limit from the Dirac-Yang-Mills theory (16).

Define the isovector source current for N classical point-like non-Abelian charges by the expression

$$\underset{\sim}{j}_\nu = \sum_{n=1}^{N} \int_{-\infty}^{\infty} \underset{\sim}{v}_{(n)\nu} \delta^{(4)}[x-\xi_{(n)}(\tau)] d\tau \ , \tag{3.18}$$

where $\xi^\mu_{(n)}(\tau)$ denotes the trajectory of the nth particle, as a function of the proper time τ, and $\underset{\sim}{v}_{(n)\mu}(\tau)$ is an isovector space-time four-vector for the nth particle which will be shown below to have the form

$$\underset{\sim}{v}^\mu_{(n)}(\tau) = \underset{\sim}{q}_{(n)}(\tau)\dot{\xi}^\mu_{(n)}(\tau) \ ,$$

with $\underset{\sim}{q}_{(n)}(\tau)$ being the SU(2) charge at the proper time τ. As a consequence of the field equation, the current in Eq. (3.18) is covariantly conserved, that is

$$\nabla^\nu \underset{\sim}{j}_\nu = \partial^\nu \underset{\sim}{j}_\nu + \underset{\sim}{b}^\nu \times \underset{\sim}{j}_\nu = 0 \ . \tag{3.19}$$

(Notice that our $\underset{\sim}{b}^\mu$ has an opposite sign of that of Drechsler and Rosenblum.)

Besides the source current $\underset{\sim}{j}_\nu$, the symmetric energy-momentum tensor of matter of the N non-Abelian charged particle of masses $m_{(n)}$ is given by

$$t^{\mu\nu} = \sum_{n=1}^{N} \int_{-\infty}^{\infty} p^{\mu\nu}_{(n)}(\tau)\delta^{(4)}[x-\xi_{(n)}(\tau)] d\tau \ . \tag{3.20}$$

$p^{\mu\nu}_{(n)}(\tau)$ will be determined as a consequence of the energy and momentum conservation. In addition, the energy-momentum of the gauge fields $\underset{\sim}{f}_{\mu\nu}$ is given by

$$T^{\mu\nu}_{(f)} = \frac{1}{4} \eta^{\mu\nu} \underset{\sim}{f}^{\kappa\lambda} \cdot \underset{\sim}{f}_{\kappa\lambda} - \underset{\sim}{f}^{\mu\kappa} \cdot \underset{\sim}{f}^\nu{}_\kappa \ . \tag{3.21}$$

It is easy to show from the field equations that

$$\partial_\mu T^{\mu\nu}_{(f)} = - \underset{\sim}{f}^{\nu\rho} \cdot \underset{\sim}{j}_\rho \ . \tag{3.22}$$

The total energy and momentum conservation for a system composed of material particles and gauge fields is expressed as

$$\partial_\mu (t^{\mu\nu} + T^{\mu\nu}_{(f)}) = 0 \ . \tag{3.23}$$

Using Eq. (3.22) this can be written in the form

$$\partial_\mu t^{\mu\nu} - \underset{\sim}{f}^{\nu\rho} \cdot \underset{\sim}{j}_\rho = 0 \ . \tag{3.24}$$

To derive the equations of motion of the particles, one uses Eqs. (3.18) and (3.20) in Eqs. (3.19) and (3.24), respectively, multiplying the first equation by an arbitrary smooth function S(x) and the second by a set of arbitrary smooth functions $S_\nu(x)$, and integrating over all space-time. One obtains

$$\sum_{n=1}^{N} \iint_{-\infty}^{\infty} \Big(\underset{\sim}{v}^\nu_{(n)} \partial_\nu \delta^{(4)}[x-\xi_{(n)}(\tau)] + \underset{\sim}{b}_\nu \times \underset{\sim}{v}^\nu_{(n)} \delta^{(4)}[x-\xi_{(n)}(\tau)] \Big) S(x) d\tau \ d^4x = 0 \ , \tag{3.25}$$

$$\sum_{n=1}^{N} \iint_{-\infty}^{\infty} \Big(p^{\mu\nu}_{(n)} \partial_\mu \delta^{(4)}[x-\xi_{(n)}(\tau)] - \underset{\sim}{f}^{\nu\rho} \cdot \underset{\sim}{v}_{(n)\rho} \delta^{(4)}[x-\xi_{(n)}(\tau)] \Big) S_\nu(x) d\tau \ d^4x = 0 \ . \tag{3.26}$$

Demanding that the functions S(x) and $S_\nu(x)$ vanish at the limits of the x and τ integrations, and performing an integration by parts on the first terms on the left-hand sides of Eqs. (3.25) and (3.26) over d^4x, one obtains

$$\sum_{n=1}^{N} \int_{-\infty}^{\infty} \Big[\underset{\sim}{v}^\nu_{(n)} \partial_\nu S - \underset{\sim}{b}_\nu \times \underset{\sim}{v}^\nu_{(n)} S \Big]_{x=\xi_{(n)}(\tau)} d\tau = 0 \ , \tag{3.27}$$

$$\sum_{n=1}^{N} \int_{-\infty}^{\infty} \Big[p^{\mu\nu}_{(n)} \partial_\mu S_\nu + \underset{\sim}{f}^{\nu\rho} \cdot \underset{\sim}{v}_{(n)\rho} S_\nu \Big]_{x=\xi_{(n)}(\tau)} d\tau = 0 \ . \tag{3.28}$$

Following Mathisson (17), one breaks $p^{\mu\nu}_{(n)}$ and $\underset{\sim}{v}^\mu_{(n)}$ into components parallel and perpendicular to the world line at the point determined by τ. To this end one introduces, besides the velocity vectors $\dot{\xi}^\mu_{(n)}(\tau) = \dot{\xi}^\mu_{(n)}$, the space-like unit vectors $n^\mu_{(n)}(\tau) = n^\mu_{(n)}$ obeying for each n

$$n^\mu_{(n)} n_{(n)\mu} = -1, \qquad \dot{\xi}^\mu_{(n)} n_{(n)\mu} = 0 \ , \tag{3.29}$$

and hence

$$\underset{\sim}{v}^\nu_{(n)} = \underset{\sim}{q}_{(n)} \dot{\xi}^\nu_{(n)} + \underset{\sim}{I}^\nu_{(n)} \ , \tag{3.30}$$

$$P^{\mu\nu}_{(n)} = m_{(n)}\dot{\xi}^{\mu}_{(n)}\dot{\xi}^{\nu}_{(n)} + a\left(n^{\mu}_{(n)}\dot{\xi}^{\nu}_{(n)} + n^{\nu}_{(n)}\dot{\xi}^{\mu}_{(n)}\right) + P^{\mu\nu}_{(n)} \ , \quad (3.31)$$

with the transverse parts $\underset{\sim}{I}^{\nu}_{(n)}$ and $P^{\mu\nu}_{(n)} = P^{\nu\mu}_{(n)}$ satisfying

$$\underset{\sim}{I}^{\nu}_{(n)}\dot{\xi}_{(n)\nu} = 0 \ , \quad (3.32)$$

$$P^{\mu\nu}_{(n)}\dot{\xi}_{(n)\nu} = 0 \ . \quad (3.33)$$

Substituting Eqs. (3.30) and (3.31) into Eqs. (3.27) and (3.28) one finds

$$\sum_{n=1}^{N}\int_{-\infty}^{\infty}\Big\{\Big[\underset{\sim}{q}_{(n)}\dot{\xi}^{\nu}_{(n)} + \underset{\sim}{I}^{\nu}_{(n)}\Big]\partial_{\nu}S$$
$$- \underset{\sim}{b}_{\nu}\times\Big[\underset{\sim}{q}_{(n)}\dot{\xi}^{\nu}_{(n)} + \underset{\sim}{I}^{\nu}_{(n)}\Big]S\Big\}_{x=\xi_{(n)}(\tau)}d\tau = 0 \ , \quad (3.34a)$$

$$\sum_{n=1}^{N}\int_{-\infty}^{\infty}\Big\{\Big[m_{(n)}\dot{\xi}^{\mu}_{(n)}\dot{\xi}^{\nu}_{(n)} + a\left(n^{\mu}_{(n)}\dot{\xi}^{\nu}_{(n)} + n^{\nu}_{(n)}\dot{\xi}^{\mu}_{(n)}\right) + P^{\mu\nu}_{(n)}\Big]\partial_{\mu}S_{\nu}$$
$$+ \underset{\sim}{f}^{\nu\rho}\cdot\Big[\underset{\sim}{q}_{(n)}\dot{\xi}_{(n)\rho} + \underset{\sim}{I}_{(n)\rho}\Big]S_{\nu}\Big\}_{x=\xi_{(n)}(\tau)}d\tau = 0 \ . \quad (3.34b)$$

Since $\dot{\xi}^{\mu}_{(n)}\partial_{\mu}$ for $x = \xi_{(n)}(\tau)$ is equal to $d/d\tau$ taken along the nth world line, one can perform an integration by parts to obtain

$$\sum_{n=1}^{N}\int_{-\infty}^{\infty}\Big\{\Big[(d/d\tau)\underset{\sim}{q}_{(n)} + \underset{\sim}{b}_{\nu}\times\underset{\sim}{q}_{(n)}\dot{\xi}^{\nu}_{(n)}\Big]S$$
$$- \Big[\underset{\sim}{I}^{\nu}_{(n)}\partial_{\nu}S - \underset{\sim}{b}_{\nu}\times\underset{\sim}{I}^{\nu}_{(n)}S\Big]\Big\}_{x=\xi_{(n)}(\tau)}d\tau = 0 \ , \quad (3.35)$$

$$\sum_{n-1}^{N}\int_{-\infty}^{\infty}\Big\{\Big[\frac{d}{d\tau}\left(m_{(n)}\dot{\xi}^{\nu}_{(n)} + an^{\nu}_{(n)}\right)\Big]S_{\nu} - \Big[an^{\mu}_{(n)}\dot{\xi}^{\nu}_{(n)} + P^{\mu\nu}_{(n)}\Big]\partial_{\mu}S_{\nu}$$
$$- \underset{\sim}{f}^{\nu\rho}\circ\Big[\underset{\sim}{q}_{(n)}\dot{\xi}_{(n)\rho} + \underset{\sim}{I}_{(n)\rho}\Big]S_{\nu}\Big\}_{x=\xi_{(n)}(\tau)}d\tau = 0 \ . \quad (3.36)$$

Due to the arbitrariness of S_{ν} and $\partial_{\mu}S_{\nu}$ as well as of S and $\partial_{\mu}S$ at $x = \xi_{(n)}(\tau)$,

$$\underset{\sim}{I}_{(n)\nu} = 0 \ , \quad (3.37)$$

and

$$an^{\mu}_{(n)}\dot{\xi}^{\nu}_{(n)} + P^{\mu\nu}_{(n)} = 0 \ . \qquad (3.38)$$

Contracting the latter equation with $\dot{\xi}_{(n)\nu}$ gives a = 0. Equation (3.38) then yields $P^{\mu\nu}_{(n)} = 0$, and the final equations are

$$(d/d\tau)\underset{\sim}{q}_{(n)} + \underset{\sim}{b}_{\nu} \times \underset{\sim}{q}_{(n)}\dot{\xi}^{\nu}_{(n)} = 0 \ , \qquad (3.39)$$

$$(d/d\tau)\left(m_{(n)}\dot{\xi}^{\nu}_{(n)}\right) = \underset{\sim}{q}_{(n)}\cdot\underset{\sim}{f}^{\nu\rho}\dot{\xi}_{(n)\rho} \ . \qquad (3.40)$$

These are Drechsler and Rosenblum's equations of motion for the non-Abelian charge. Equation (3.39) is the gauge covariant charge precession equation, and Eq. (3.40) represents the Lorentz-type equation for the trajectory of the motion. For a single charge, dropping the subscript (n), one gets

$$\frac{d\underset{\sim}{q}}{d\tau} + \dot{\xi}^{\nu}\underset{\sim}{b}_{\nu} \times \underset{\sim}{q} = 0 \ , \qquad (3.41)$$

$$\frac{d(m\dot{\xi}^{\mu})}{d\tau} = \underset{\sim}{q}\cdot\underset{\sim}{f}^{\mu\rho}\dot{\xi}_{\rho} \ . \qquad (3.42)$$

Equations (3.41) and (3.42) are identical to those of Wong, Eqs. (3.15) and (3.14), if we take $\underset{\sim}{I} = \underset{\sim}{q}$ and g is reinstated.

4. SOME SOLUTIONS OF THE EQUATIONS OF MOTION

In the last section we derived the equations of motion of a classical particle in both the Minkowskian and the internal spaces. We now solve these equations when the fields and potentials are given by Eqs. (2.17) and (2.18). A straightforward calculation then yields:

$$m\ddot{\xi}^{0} = 0 \ , \qquad (4.1a)$$

$$m\ddot{\xi}^{1} = 0 \ , \qquad (4.1b)$$

$$m\ddot{\xi}^{2} = -\ \underset{\sim}{\alpha}\cdot\underset{\sim}{I}\dot{\xi}^{3} \ , \qquad (4.1c)$$

$$m\ddot{\xi}^{3} = \underset{\sim}{\alpha}\cdot\underset{\sim}{I}\dot{\xi}^{2} \ , \qquad (4.1d)$$

and

$$\dot{I}_k + g\varepsilon_{ijk}b_iI_j = 0 \ , \qquad (4.2)$$

with $j^{\mu} = 0$. In Eq. (4.2), $b_i = b^{\mu}_{i}\dot{\xi}_{\mu}$, and $\underset{\sim}{\alpha}$ is given by Eq. (2.5).

Multiplying Eq. (4.2) by I_k we find that $\underset{\sim}{I}\cdot\dot{\underset{\sim}{I}} = 0$ and therefore

$$d\underset{\sim}{I}^{2}/d\tau = 0 \ . \qquad (4.3)$$

Consequently

$$\underset{\sim}{I} = R\underset{\sim}{n} , \tag{4.4}$$

where R is independent of τ, and $\underset{\sim}{n}$ is the unit vector given by

$$\underset{\sim}{n} = (\sin\theta \cos\varphi, \quad \sin\theta \sin\varphi, \quad \cos\theta) . \tag{4.5}$$

Thus the particle moves in a circle with radius R in the isospin space, and we have

$$\underset{\sim}{\alpha}\cdot\underset{\sim}{I} = KR \sin\theta \cos(\omega-\varphi) .$$

To investigate the motion of the particle in the ordinary space-time we have to solve Eqs. (4.1). The first two equations are trivial and their solutions are given by

$$\xi^0 = C_1\tau + C_2 , \tag{4.6a}$$

$$\xi^1 = C_3\tau + C_4 , \tag{4.6b}$$

where C_i are arbitrary constants. Differentiating Eqs. (4.1c) and (4.1d) with respect to τ, and using the following substitutions

$$\dot{\xi}^2 = \eta , \quad \dot{\xi}^3 = \zeta , \tag{4.7}$$

then we get

$$\ddot{\eta} - \frac{\dot{f}}{f}\dot{\eta} + f^2\eta = 0 , \tag{4.8a}$$

$$\ddot{\zeta} - \frac{\dot{f}}{f}\dot{\zeta} + f^2\zeta = 0 , \tag{4.8b}$$

where $f = (1/m)\underset{\sim}{\alpha}\cdot\underset{\sim}{I}$ is a function of the space-time coordinates x.

Introducing now the new parameter

$$s = \int f \, d\tau , \tag{4.9}$$

then Eqs. (4.8) are reduced to the simplified forms

$$\frac{d^2\eta}{ds^2} + \eta = 0 , \tag{4.10a}$$

$$\frac{d^2\zeta}{ds^2} + \zeta = 0 . \tag{4.10b}$$

The solutions of these equations are then given by

$$\eta = E_1 \cos s + E_2 \sin s , \tag{4.11a}$$

$$\zeta = -E_2 \cos s + E_1 \sin s , \tag{4.11b}$$

where E_1 and E_2 are arbitrary constants. Recalling that $s = \int f d\tau$, $\eta = \dot{\xi}^2$ and $\zeta = \dot{\xi}^3$, one sees that the velocities of the classical particle in these two coordinates behave like the coordinates of harmonic oscillators when the parameter s is used.

To find ξ^2 and ξ^3 we have to integrate Eqs. (4.11) with respect to τ. We then obtain, using $d\tau = \dot{s}^{-1} ds$,

$$\xi^2 = \int \frac{\eta(s)}{\dot{s}} ds + \xi_0^2 , \tag{4.12a}$$

$$\xi^3 = \int \frac{\zeta(s)}{\dot{s}} ds + \xi_0^3 , \tag{4.12b}$$

where ξ_0^2 and ξ_0^3 are constants.

The integrals in the above equations are functions of the parameter s which, by turn, depends on the function f. We now consider two particular cases of f.

1) Let us first take

$$f = \Omega , \tag{4.13}$$

where Ω is a constant, and therefore the angle between the two isospin vectors $\underset{\sim}{I}$ and $\underset{\sim}{\alpha}$ is constant. We obtain

$$\eta = E_1 \cos \Omega\tau + E_2 \sin \Omega\tau , \tag{4.14a}$$

$$\zeta = -E_2 \cos \Omega\tau + E_1 \sin \Omega\tau . \tag{4.14b}$$

Hence ξ^2 and ξ^3 are given by

$$\xi^2 = \frac{1}{\Omega} [E_1 \sin \Omega\tau - E_2 \cos \Omega\tau] + \xi_0^2 , \tag{4.15a}$$

$$\xi^3 = \frac{-1}{\Omega} [E_2 \sin \Omega\tau + E_1 \cos \Omega\tau] + \xi_0^3 . \tag{4.15b}$$

In this case, we see that the motion of the classical particle is harmonic with a constant frequency Ω.

2) We now take

$$\frac{\dot{f}}{f} = \kappa , \tag{4.16}$$

where κ is a constant. We then obtain

$$f = f_0 e^{\kappa\tau} , \quad f_0 = \text{constant}, \tag{4.17}$$

and

$$s = \kappa^{-1} f . \tag{4.18}$$

REFERENCES

1. Castillejo, L., Kugler, M. and Roskies, R.Z. 1979, Phys. Rev. D 19, pp. 1782-1790.
2. Carmeli, M., Charach, Ch. and Kaye, M. 1978, Nuovo Cimento 45B, pp. 310-334.
3. Carmeli, M. 1978, Phys. Lett. 77B, pp. 188-192.
4. Carmeli, M. 1982, *Classical Fields*, Wiley-Interscience, New York, Chap. 9.
5. Wang, L.L. and Yang, C.N. 1978, Phys. Rev. D 17, pp. 2687-2694.
6. Newman, E.T. 1978, Phys. Rev. D 18, pp. 2901-2908.
7. Newman, E.T. 1980, Phys. Rev. D 22, pp. 3023-3033.
8. Newman, E.T. and Penrose, R. 1962, J. Math. Phys. 3, pp. 566-578.
9. Carmeli, M. 1977, *Group Theory and General Relativity*, McGraw-Hill, New York, Appendix D.
10. Yang, C.N. and Mills, R.L. 1954, Phys. Rev. 96, pp. 191-195.
11. Kinnersley, W. 1969, J. Math. Phys. 10, pp. 1195-1214.
12. Carmeli, M. 1977, Phys. Lett. 68B, pp. 463-465.
13. Altamirano, L. and Villarroel, D. 1981, Phys. Rev. D 24, pp. 3118-3124.
14. Castillejo, L. and Kugler, M. 1981, Phys. Rev. D 24, pp. 2626-2635.
15. Wong, S.K. 1970, Nuovo Cimento 65A, pp. 689-694.
16. Drechsler, W. and Rosenblum, A. 1981, Phys. Lett. 106B, pp. 81-87.
17. Mathisson, M. 1940, Proc. Cambridge Philos. Soc. 36, pp. 331-340.
18. Havas, P. 1979, in: *Isolated Gravitating Systems in General Relativity*, J. Ehlers, Editor, North-Holland, Amsterdam.
19. Landau, L.D. and Lifshitz, E.M. 1975, *The Classical Theory of Fields*, Pergamon, New York, Sect. 21.

A NORMAL FORM FOR THE MOMENT MAP

Victor Guillemin and Shlomo Sternberg

1. Let M be a connected symplectic manifold with symplectic form ω. A vector field, ζ, on M is called <u>Hamiltonian</u> if there is a function, f, on M such that

$$i(\zeta)\omega = df \; . \tag{1.1}$$

Since ω is non-degenerate, df is determined by this equation and hence, since M is connected, f is determined up to a constant, and, in any event, f determines ζ uniquely, so we may write ζ_f for the vector field determined by f. The set of all Hamiltonian vector fields is a Lie algebra under the usual Lie bracket of vector fields. The space of all smooth functions on M is a Lie algebra under Poisson bracket defined by

$$\{f_1,f_2\} = -D_{\zeta_{f_1}} f_2$$

where D_ζ denotes Lie derivative with respect to ζ.

Let $F(M)$ denote the space of smooth functions on M, and Ham(M) the space of Hamiltonian vector fields on M. The map from $F(M)$ to $\mathrm{Ham}(M)$ sending f into ζ_f is a homomorphism of Lie algebras and we have the exact sequence

$$0 \to R \to F(M) \to \mathrm{Ham}(M) \to 0$$

where R denotes the trivial Lie algebra of constant functions.

For any vector field ζ and any differential form ω, the fundamental formula of differential calculus says that

$$D_\zeta\omega = i(\zeta)d\omega + di(\zeta)\omega.$$

If ω is symplectic so that $d\omega = 0$ this becomes

S. Sternberg (ed.), Differential Geometric Methods in Mathematical Physics, 161–175.

$$D_\zeta\omega = di(\zeta)\omega .$$

Thus if ζ is an infinitesimal symplectic transformation, so that $D_\zeta\omega = 0$ then

$$di(\zeta)\omega = 0$$

which is a weaker condition than (1.1) (and equivalent to it if $H^1(M) = \{0\}$.)

Let G be a Lie group with Lie algebra g acting on M as symplectic diffeomorphisms. Each $\xi \in g$ gives rise to a vector field ξ_M so that we have a homomorphism

$$g \to \mathrm{Symp}(M)$$

where Symp(M) denotes the Lie algebra of infinitesimal symplectic vector fields of M.

We thus have the diagram of Lie algebra homomorphisms

$$\begin{array}{ccccc} F(M) & \to & \mathrm{Ham}(M) & \to & \mathrm{Symp}(M) \\ & & & \nearrow & \\ & & g & & \end{array}$$

The action is called Hamiltonian if we are given a Lie algebra homomorphism $g \to F(M)$ which is a G morphism, so that the diagram

$$\begin{array}{ccccc} F(M) & \to & \mathrm{Ham}(M) & \to & \mathrm{Symp}(M) \\ & \nwarrow & & \nearrow & \\ & & g & & \end{array}$$

commutes. Thus, for each $\xi \in g$ the vector field ξ_M is Hamiltonian so that we have a function f_ζ such that

$$i(\xi)\omega = df_\zeta$$

and the map $\xi \to f_\xi$ is a Lie algebra homomorphism. In particular, the dependence of f_ξ on ξ is linear. Thus we have a map $\Phi: M \to g^*$ called the <u>moment map</u> defined by

$$\langle\Phi,\xi\rangle = f_\xi$$

where $\langle,\rangle$ denotes the pairing between g^* and g. Thus

$$d\langle\Phi,\xi\rangle = i(\xi_M)\omega$$

for each $\xi \in g$. In this last equation, ξ is a constant (an element of g). We can thus write its left hand side as $\langle d\Phi,\xi\rangle$. At each point p of M we can think of $d\Phi_p$ as a map from TM_p to g^*. Evaluating on a vector $v \in TM_p$ gives

$$\langle d\Phi_p(v),\xi\rangle = (\xi,v)_p$$

where $(\, , \,)_p$ denotes the symplectic bilinear form on TM_p given by evaluating ω at p.

Another fact about the moment map that we will use is that it is G equivariant. That is,

$$\Phi(am) = a\Phi(m)$$

where the action on the right-hand side of this equation is the coadjoint representation of G on g^* - the contragredient to the adjoint representation.

In this paper we shall present a local normal form for Hamiltonian group actions and moment maps, when the group G is compact.

2. There are three cases of Hamiltonian group actions where the computation of the moment map is particularly simple:

1) The case of a linear symplectic action. If we are given a linear symplectic action of G on a symplectic vector space, the corresponding moment map $\Phi_V\colon V \to g^*$ is quadratic. It is given by

$$\langle\Phi_V(v),\xi\rangle = \tfrac{1}{2}(\xi v,v) \tag{2.1}$$

where $\xi \in g$, $v \in V$, and ξv denotes the action of ξ or v, $(\, , \,)$ is the symplectic scalar product, and $\langle,\rangle$ is the pairing between g* and g. This can be checked directly:

$$d\langle\Phi_V(\cdot),\xi\rangle_v(\omega) = \tfrac{1}{2}(\xi v,\omega) + \tfrac{1}{2}(\xi\omega,v) = (\xi v,\omega)$$

since $(\xi\omega,v) + (\omega,\xi v) = 0$ (as ξ is infinitesimally symplectic) and (,) is antisymmetric. This is the defining property of the moment map. A direct check shows the (1.1) is equivariant and hence is indeed a moment map.

2) Let G act on itself by left multiplication and consider the induced action on T^*G. We may use a left invariant identification of T^*G with $G\times g^*$. (This means that we are using right multiplication by c to identify T^*G_c with T^*G_e). Then the induced action is given by

$$\ell_a(c,\alpha) = (ac,\alpha).$$

Under this identification, the Lie algebra g is identified as the space of left invariant vector fields. In particular, the identification of T^*G_e with g^* is exactly the moment map, i.e.,

$$\Phi_\ell(e,\alpha) = \alpha$$

(where ℓ stands for left). Now the moment map is always equivariant:

$$\Phi(cm) = c\Phi(m).$$

Since $(c,\alpha) = c(e,\alpha)$ we see that

$$\Phi_\ell(c,\alpha) = c\alpha .$$

3) We define the right multiplication action of G on itself by

$$r_b c = cb^{-1}$$

since we want this to be a left action, and consider the induced action on T^*G. Under the left invariant identification used in 2) this becomes

$$r_b(c,\alpha) = (cb^{-1}, b\alpha).$$

The element ξ of g generates right multiplication by exp-tξ whose infinitesimal generator is the left invariant vector field $-\xi$. So there is a minus sign and hence

$$\Phi_r(e,\alpha) = -\alpha .$$

Since $r_{c^{-1}}(e,c\alpha) = (c,\alpha)$ equivariance implies that

$$\Phi_r(c,\alpha) = -\alpha .$$

In this paper we shall show that if G is a compact connected Lie group then the most general moment map can be expressed locally as a contribution of the above three types.

3. In fact, let M be a Hamiltonian G space, and let p be a point of M. We shall show that M is completely determined in a G invariant neighbourhood of p (as a Hamiltonian G space) by the following data

1) $\alpha = \Phi(p)$
2) the isotropy group G_p
3) the linear representation of G_p on the tangent space TM_p.

We will first prove this result under the auxiliary hypothesis that α is left fixed by all of G. For example, this will be the situation if $\alpha = 0$, an important special case.

By the definition of the moment map, we have

$$d\langle\Phi,\xi\rangle = i(\xi_M)\omega$$

where ξ_M is the vector field on M corresponding to $\xi \in g$. We can write this last equation as

$$\langle d\phi, \xi\rangle = i(\xi_M)\omega \ .$$

Evaluating this at a vector v in TM_p, and writing $(\ ,\ )_p$ for the form ω_p we can write this last equation as

$$(3.1) \qquad \langle d\Phi_p(v), \xi\rangle = (\xi_M(p), v)_p \ .$$

We will first apply this to $v = \eta_M(p)$. The equivariance of ϕ implies that

$$d\Phi_p(\eta_M(P)) = \eta \cdot \alpha$$

and our hypothesis implies that $\eta \cdot \alpha = 0$. Thus

$$(\xi_M(p),\ \eta_M(p))_p = 0$$

so the space $g_M(p) \subset TM_p$ consisting of all $\xi_M(p)$, $\xi \in g$ is isotropic, i.e.,

$$g_M(p) \subset g_M(p)^{\perp} \ .$$

This implies that the orbit of p

$$Z = G \cdot p$$

is an isotropic submanifold of M.

We now recall some important facts about isotropic submanifolds of symplectic manifolds due to Weinstein:

Let M be a symplectic manifold and $i: X \to M$ an isotropic embedding of some other manifold, X, into M. This means that i is an embedding and that at each $x \in X$, $di_x(TX_x)$ is an isotropic subspace of $TM_{i(x)}$. Thus

$$di_x(TX_x) \subset (di_x(TX_x))^{\perp} \subset TM_{i(x)} \ .$$

We shall write this more simply as

$$TX \subset TX^{\perp} \subset TM_{|X} \quad .$$

We can form the vector bundle

$$E = (TX)^{\perp}/TX$$

which Weinstein calls the symplectic normal bundle (to (X,i)). If $i_1: X \to M_1$ and $i_2: X \to M_2$ are two isotropic embeddings and if f is a symplectic diffeomorphism of some neighbourhood of $i_1(X)$ into M_2 satisfying

$$(3.2) \qquad i_2 = f \cdot i_1$$

then f defines a symplectic bundle isomorphism, $L_f: E_1 \to E_2$ (where E_1 and E_2 denote the corresponding symplectic normal bundles). The map sending f to L_f is functorial in the obvious sense.

In [3], Weinstein proves the following:

Isotropic embedding theorem. Any isotropic embedding, $i: X \to M$ determines a symplectic normal bundle $E \to X$ where $E = TX^{\perp}/TX$. If f is a symplectic diffeomorphism defined on some neighborhood of $i_1(X)$ satisfying (3.2) where i_1 and i_2 are isotropic embeddings of X, then f induces a symplectic isomorphism, $L_f: E_1 \to E_2$ of the corresponding symplectic normal bundles. Conversely, given any symplectic isomorphism, $L: E_1 \to E_2$, there exists an f with $L = L_f$. Given any symplectic vector bundle $E \to X$ there exists an isotropic embedding of X whose symplectic normal bundle is E.

An examination of Weinstein's proof shows that if we are given a compact group G, all the assertions are valid within the category of equivariant G maps.

In our application, this theorem implies that a symplectic neighbourhood of Z is completely determined by the symplectic normal bundle $E \to Z$. As G acts transitively on Z, the vector bundle E is a homogeneous G bundle and so is completely determined by the action of G_p on $V = E_p$, the fiber over p, but $V = TZ_p^{\perp}/TZ_p$. Hence the representation of G_p on TM_p determines the action of G on E and hence the action of G on a symplectic neighbourhood of Z. We have shown that (up to a symplectic G morphism) M is determined, as a <u>symplectic</u> G space near p by the representation of G_p on TM_p. As a <u>Hamiltonian</u> G space, G is determined by its symplectic action, with the possible indeterminacy of an additive constant in the moment map. This possible indeterminacy is removed by knowing that $\Phi(p) = \alpha$.

We can find a convenient normal form for M and Φ near Z by making a slight modification in Weinstein's proof of the existence of an isotropic embedding: Let $G \to Z = G/G_p$ be the principal bundle with structure group G_p. For convenience let us write $H = G_p$. Let

$$i = h \to g$$

be the corresponding injection of Lie algebras and

$$i^*: \quad g^* \to h^*$$

the dual projection. Thus E can be regarded as the associated bundle to G corresponding to the representation of H on V:

$$E = G \times V/H.$$

Then, cf. Weinstein [ ], Z has an isotropic embedding in T^*G_V, the Marsden-Weinstein reduced space of T*G at the Hamiltonian G space, V. Indeed

$$\overline{\Phi}: \quad T^*G^- \times V \to h^*$$

is given by

(3.3) $$\overline{\Phi} = \Phi_V - \Phi_G$$

where Φ_V is the moment map for the linear H action on V (thought of as being defined on the product space $(^*G^- \times V)$ and Φ_G is the moment map for the right action of H on G. Then $\overline{\Phi}^{-1}(0)$ is a coisotropic submanifold of $T^*G^- \times V$ and

$$\overline{\Phi}^{-1}(0)/H = T^*G_V$$

is a symplectic manifold.

Contained in $\overline{\Phi}^{-1}(0)$ is the set of all points of the form $(z,0)$ where $z \in T^*P$ satisfies $\Phi_G(z) = 0$. Now $\Phi(z) = 0$ if and only if z vanishes on all vertical tangent vectors. Thus $z = d\pi_p^* \ell$ for some $\ell \in T^*Z_{\pi(p)}$. So $\overline{\Phi}^{-1}(0)/H$ is canonically identified with T^*Z, and it is easy to check that with this identification, T^*Z together with its standard symplectic structure is a symplectic submanifold of T^*G. But Z is embedded in T^*Z as the zero section, and this is clearly an isotropic embedding. A choice of connection shows that the full Marsden-Weinstein reduced space T^*G_V can be identified with the pullback, $E\#$ of E to T^*Z. From this we see that the symplectic normal bundle to Z in T^*G_V can be identified with our original symplectic vector bundle, E. This gives an isotropic embedding of Z in T^*G_V and thus, by the uniqueness part of the isotropic embedding theorem, our original manifold M can be identified with T^*G_V near Z.

Let us examine what this Marsden-Weinstein reduced space (and its G moment map) looks like in terms of the left invariant identification of T^*G with $G \times g^*$. The moment map, $\Phi_H: T^*G \to h^*$ for the right action of H on T^*G is given, in terms of this identification by

$$\Phi_H(c,\beta) = -i^*\beta \qquad c \in g \text{ and } \beta \in g^*.$$

The moment map $\Phi_V: V \to h^*$ is the quadratic map given by

$$\langle\Phi_V(v),\eta\rangle = \tfrac{1}{2}(\eta v,v)_V \qquad v \in V,\ \eta \in h,$$

where $(\ ,\ )_V$ denotes the symplectic form on V.

Thus $\overline{\Phi}^{-1}(0)$ consists of all (c,β,v) such that

$$i^*\beta = -\Phi_V(v)\ . \tag{3.4}$$

The group H acts on $\overline{\Phi}^{-1}(0)$ by

$$r_b(c,\beta,v) = (cb^{-1},b\cdot\beta,bv), \qquad b \in H\ .$$

The space T^*G_V is the quotient space of $\overline{\Phi}^{-1}(0)$ by this H action. Let $V \subset g^*\times V$ be the quadratic variety consisting of all pairs (β,v) satisfying (*). We thus see that T^*G_V can be identified as the associated bundle, $V(G)$. The symplectic action of G on T^*G_V is that induced from left multiplication on G and the trivial action on V:

$$\ell_a(c,\beta,v) = (ac,\beta,v),$$

The moment map coming from this co-adjoint action is

$$\Phi_G(c,\beta,v) = c\cdot\beta\ .$$

This is clearly constant on H orbits and hence descends to give a moment map, Φ': $T^*G_V \to g^*$. Now the symplectic structure and G action is determined by the isotropic embedding theorem, but the moment map (when it exists) is only determined up to a constant. The map Φ' has the property that it sends Z into 0. Our original moment map sent Z into α. Thus we have proved that a normal form for our moment map is given in terms of the above description of T^*G_V is given by

$$\Phi([(c,\beta,v)]) = \alpha + c\cdot\beta\ ,$$

where (*) holds and where [ ] denotes the equivalence class, that is, the quotient modulo the H action.

4. We will now prove the theorem in the general case. For this we will need some preliminary facts about the moment map that are proved in [2].

Let G be a connected Lie group with a Hamiltonian action on M and with moment map Φ: $M \to g^*$. Let p be a point of M and let $\alpha = \Phi(p) \in g^*$. Let Y be a submanifold of g^* passing through α which is transversal to $\mathcal{O} = G\cdot\alpha$ at α in the sense that

(4.1) $$g^* = T\mathcal{O}_\alpha \oplus TY_\alpha .$$

Then, cf. [2],

Prop. 4.1. <u>There is a neighbourhood</u> U <u>of</u> p <u>such that</u> $\Phi^{-1}(Y) \cap U$ <u>is a symplectic submanifold of</u> U.

In case G is compact, we can make a particularly nice choice of Y: Let $K \subset G$ be the stabilizer group of α.

It is clear that K is closed and it is known that K is connected. Let L be the center of K, and let k and ℓ denote the corresponding Lie algebras. We can consider the following subspaces of g^*:

$k^{\#}$ = the set of all elements of g^* stabilized by L

and

$\ell^{\#}$ = the set of all elements of g^* stabilized by K.

Clearly

$$\ell^{\#} \subset k^{\#}$$

and

$$\alpha \in \ell^{\#} .$$

It is also clear that K preserves $k^{\#}$. In [ ] the following facts are proved:

Prop. 4.2

(1) The orbit $\mathcal{O}$ through α intersects $k^{\#}$ transversally at α and at this intersection (4.1) is satisfied, with $Y = k^{\#}$.

(2) For points $\beta \in k^{\#}$ near α the stabilizer of β in G is contained in K.

(3) The canonical projection, $g^* \to k^*$ maps $k^{\#}$ bijectively onto k^*.

(4) The canonical projection, $g^* \to M^*$ bijectively onto M^*.

(5) Every co-adjoint orbit in g^* intersects $k^{\#}$ in a finite number of K orbits.

(6) There is a neighborhood, $B_\varepsilon(\alpha)$ of α in $k^{\#}$ such that every co-adjoint orbit intersections $B_\varepsilon(\alpha)$ in exactly one K orbit.

By Prop. 4.1 we can find a (K invariant) neighbourhood U of $\Phi^{-1}(\alpha)$ such that

$$W = \Phi^{-1}(B_\varepsilon(\alpha)) \cap U$$

is a symplectic submanifold of M. We note the following properties of W:

Prop. 4.3 a) W is K invariant

b) the action of K on W is Hamiltonian and the associated moment map is the map $\Phi\colon W \to R^{\#}$ composed with the identification of $k^{\#}$ with k^*

c) the intersection of any G orbit in M with W is a K orbit.

<u>Proof.</u> Parts a) and b) are obvious. To prove c), let p_1 and p be two points of W with $p_1 = ap_2$, $a \in G$.

Then $\Phi_1(p)$ and $\Phi_2(p)$ both lie in $B_\varepsilon(\alpha)$ and hence, by part of Prop. 3.2, we can find a $b \in K$ such that $b\Phi(p_1) = \Phi(p_2)$. Replacing p_1 by bp_1 we may assume that $\Phi(p_2) = \Phi(p_2)$. Thus $p_1 = ap_2$ and $a\Phi_1(p_1) = \Phi_1(p_1) = \Phi(p_2)$. By part 2) of Prop. 3.2 this implies that $a \in K$.

Let M_0 be the union of all points whose G orbits intersect W. Then M_0 is a G invariant open set containing p. We shall show how to reconstruct M_0 from W in a canonical way. Since we can apply the results of section 2 to W and K this will then allow the reconstruction of M_0 from the linear data.

As in section two, let Φ_k: $T^*G \to g^*$ denote the moment map for the right action and consider $Z = \Phi_k^{-1}(k^\#)$. Under the left invariant identification of T^*G with $G \times g^*$ the moment map Φ_k is, as we have seen in section 2, just the negative of the projection onto the second factor. Thus

$$Z = G \times k^\#.$$

By Prop. 4.1 there will be some neighbourhood $B_G(\alpha)$ of α in $k^\#$ such that

$$Z_\alpha = G \times B_\varepsilon$$

is a symplectic submanifold. Notice that Z_α is invariant under the left multiplication action of G and under the right multiplication action of K. Thus it is $G \times K$ invariant. We can now form

$$Z_\alpha \times W^-$$

which is a Hamiltonian K space, and reduce at $0 \in k^*$. Denote the resulting Hamiltonian G space by M_1.

Prop. 4.3 M_0 and M_1 are isomorphic as Hamiltonian G spaces.

<u>Proof</u>. As an abstract set

$$M_1 = (G \times W)/K.$$

Map W into M_1 by the mapping

$$i\colon W \to M_1, \qquad w \to \text{equivalence class of } (e,w).$$

It is easy to check that i is an imbedding and is K-equivariant. There is a unique way of extending i to a G-equivariant map of M_0 onto M_1. Namely, if $x \in M_0$, pick elements $g \in G$ and $w \in W$ such that $x = gw$ and set $\sigma(x) = gi(w)$. Let us show that this unambiguously defines a map $\sigma\colon M_0 \to M_1$. Suppose we also have $x = g_1 w_1$. Then $g^{-1}g_1 w_1$ and w_1 are on the same G orbit in W, so they are on the same K orbit; that is, there exists $k_0 \in K$ such that $k_0 g^{-1} g$ stabilizes w_1. This implies that $k_0 g^{-1} g$ stabilizes the point $\Phi(w_1)$ in $B_\varepsilon(\alpha)$, and, therefore, has to be in K. Thus there exists an element $k \in K$ such that

$$g_1 = gk.$$

In particular, $g_1 w_1 = gkw_1 = gw$, or

$$kw_1 = w.$$

Since i is k-equivariant, $i(w) = ki(w_1)$, so

$$gi(w) = gki(w_1) = g_1 i(w_1).$$

This proves that σ is well defined. Since σ is smooth on W and G-equivariant, it is smooth everywhere. It is easy to check that it is a symplectomorphism at all points w W and so, by G-equivariance, at all points of M_0.

We have now seen how the Hamiltonian G space, M_0, can be reconstructed from the Hamiltonian K space, W. But, for $p \in W$ and $\alpha = \Phi(p)$, K leaves α fixed. We are thus back in the situation of the beginning of this section: W, as a Hamiltonian K space, can be reconstructed from α and the linear isotropy representation of K on TW_p. We have thus completed the proof of the main theorem of this paper.

Theorem. Let G be a compact Lie group. Let M be a Hamiltonian G space and $p \in M$. Then M is completely determined near p (as a Hamiltonian G space) by $\Phi(p)$, the isotropy group, G_p, and the representation of G_p on TM_p.

REFERENCES

[1] Guillemin, V. and S. Sternberg: 1980, 'The moment map and collective motion', Annals of Physics 127, 220-253.

[2] ______________________________: (to appear), 'Multiplicity free spaces', J. Diff. Geom.

[3] Weinstein, A.: 1981, 'Neighborhood classification of isotropic embedding', J. Diff. Geom. 16, 125-128.

PLASMA KINETIC THEORY AND DIFFERENTIAL GEOMETRY
(A POISSON ALGEBRA FOR THE MAXWELL-KLIMONTOVICH SYSTEM)

Meinhard E. Mayer

Department of Physics, University of California
Irvine, California, 92717, USA

ABSTRACT. The Poisson algebra for the coupled Maxwell-Klimontovich equations of plasma kinetic theory is constructed without use of potentials by means of an extension of the Born-Infeld approach, as well as by the use of the Souriau-Sternberg momentum mapping. The full Poincaré invariance follows from the symmetric energy-momentum tensor satisfying the Dirac-Schwinger Poisson bracket relations. These ideas could be extended to quantum theory, and to nonabelian gauge theories such as the Yang-Mills theory.

1. INTRODUCTION

In spite of the growing interest in plasma kinetic theory, related to the rapid development of fusion technology, on the one hand, and of plasma astrophysics on the other, until recently not much progress has been made in this field since the early developments due to Landau, Vlasov, Klimontovich, Lenard, Rostoker, Rosenbluth, and others. Perturbative calculations have been plagued by secular divergences, the need for renormalization, etc. Only recently have plasma theorists adapted "Lie operator techniques" from celestial mechanics, and that has led to a systematic search for canonical structures in plasma kinetic theory, mainly by the school of Allan Kaufman at Berkeley. A significant step forward was made in Littlejohn's thesis (Littlejohn [1981]), when it became apparent that one must look for a symplectic structure (or a Poisson algebra), and various such structures were proposed recently, either on the basis of educated guesses (Morrison [1981]) or by applying the general theory of Poisson structures (Marsden and Weinstein [1981], Morrison and

S. Sternberg (ed.), Differential Geometric Methods in Mathematical Physics, 177–196.

Weinstein [1982]). In the meantime, John Hubbard, in his thesis work with me at U. C. Irvine, has developed a general approach, partly in collaboration with Iwo Bialynicki-Birula [1982] based implicitly on the Souriau-Sternberg idea of describing the interactions of charged particles with electromagnetic fields by modifying the symplectic structure (either the canonical two-form, or the fundamental Poisson bracket between the kinetic momenta of the particles) by adding the electromagnetic field two-form. It turned out to be possible to give a complete Poisson structure for the fully relativistic Maxwell-Klimontovich system of equations, such that the interaction between particles and electromagnetic field resides in the Poisson brackets, and the energy-momentum tensor (or the Hamiltonian density) is just the sum of those of the free particles and of the free electromagnetic field. This makes the relativistic invariance particularly transparent, and if one takes the nonrelativistic limit, leads to the Maxwell-Vlasov or Poisson-Vlasov symplectic structures previously derived by Marsden and Weinstein from their reduction technique. During the Conference Shlomo Sternberg explained to me how the Poisson structure can be constructed from general principles, and this alternative derivation is described briefly in Section 5 below.

In this paper we show how this Poisson structure is obtained in two ways: first by means of the "physical" reasoning used in Hubbard's thesis (this is preceded by an overview of the fundamental assumptions of plasma theory, for the benefit of the mathematicians in the audience) and then very quickly, by means of the use of the momentum map. In order to make the paper self-contained I have included a brief section devoted to the momentum map and Poisson structures.

The paper is organized as follows: Section 2 contains a short exposition of plasma theory for mathematicians in a very rudimentary form. Section 3 is devoted to a heuristic construction of the Poisson bracket structure for the electromagnetic field interacting with the Klimontovich distribution function. Section 4 gives a overview of Poisson structures and momentum maps (it is included for the benefit of physicists, who may not be too familiar with this language). Section 5 is devoted to rederiving the Poisson structure on the basis of the general formalism. Possible extensions and applications are discussed in Section 6. The Bibliography contains references to several papers which are not quoted directly in the text, as well as to review papers, where additional references can be found. No attempt has been made to give a complete list of references, and apologies are extended to all those who have not been quoted either out of my ignorance or oversight.

2. PLASMA PHYSICS FOR MATHEMATICIANS

This section contains a brief overview of the fundamental concepts of plasma theory, written with the mathematician in mind. Physicists may find it oversimplified, and it is obvious that the whole picture cannot be presented in a few pages. A plasma - the fourth state of matter - consists of a neutral, completely ionized gas, i.e., a gas whose neutral atoms have dissociated into electrons (of charge -e) and different species of positive ions (of charges q = Ze - we can use Z as the label of the species, but for simplicity one may assume that there are just electrons and one species of identical ions present). Each component is characterized by its density, i.e., the number of electrons per unit volume, and the number of ions per unit volume (if Z = 1 these two numbers are equal, which we assume for mathematical simplicity; the density is usually denoted by $n(\mathbf{x}, t)$ - a function of the point and time).

Plasmas can be described mathematically in various approximations; the best known are the magnetohydrodynamic description, where n is considered a smooth function, giving rise to a charge density ρ and a current density $\mathbf{j}$, and these obey hydrodynamic equations, with the forces supplied by the macroscopic electromagnetic Lorentz force on a charge-current element, and the charges and currents supplying the right-hand sides of the macroscopic Maxwell equations

$$dF = 0, \quad d{*G} = {*J}, \tag{2.1}$$

involving the two-form $F = (\mathbf{E}, \mathbf{B})$ (Λ stands for exterior product)

$$F = E_i dx^i \Lambda\, dt + \frac{1}{2} B_{ij} dx^i \Lambda\, dx^j \tag{2.1a}$$

and the dual two-form (or the appropriate Weyl-density)

$$G = -H_i dx^i \Lambda\, dt + {}^1\!/_2\, D_{ij} dx^i \Lambda\, dx^j\ . \tag{2.1b}$$

(summation over repeated latin indices is from 1 to 3; $\mathbf{E}$ and H are treated as 1-forms in 3 dimensions, whereas $\mathbf{B}$ and D are considered as two-forms); in Eq. (2.1) $*J$ is the 3-form dual to the current-density 4-vector $(\rho, \mathbf{j})$:

$$\begin{aligned} {*J} = \rho\, dx^1 \Lambda\, dx^2 \Lambda\, dx^3 + j_1 dx^2 \Lambda\, dx^3 \Lambda\, dt\, + \\ + j_2 dx^1 \Lambda\, dx^3 \Lambda\, dt + j_3 dx^1 \Lambda\, dx^2 \Lambda\, dt\ . \end{aligned} \tag{2.1c}$$

Current-conservation $d{*J} = 0$ is, of course, automatic. The fluid equations for J can be written either as an Euler equation or as Navier-Stokes equations.

Another approximation, with which I shall be concerned more in this talk, is the kinetic theory approximation, corresponding to that of an almost ideal gas (collisions are negligible, the density is so low at a given temperature that the mean kinetic energy is much larger than the electrostatic potential energy corresponding to the mean separation of particles). Here the particles are described by a density in the six-dimensional phase space of coordinates and 3-momenta, as well as of the time $f = f(\mathbf{x}, \mathbf{p}, t)$ satisfying a kinetic equation of the type (essentially a Liouville type phase-volume conservation equation)

$$\partial_t f + (\mathbf{p}/m)\cdot\partial_{\mathbf{p}} f + q(\mathbf{E} + \mathbf{v} \times \mathbf{E})\cdot\partial_{\mathbf{x}} f = 0 \quad . \tag{2.2}$$

The subscripts denote "partial differentiation with respect to vectors", i.e., a sum over components of the appropriate coefficients) and the term in parentheses describes the Lorentz force, i.e., the time derivative of the momentum **p.** I limit myself to the Klimontovich form of f,

$$f(\mathbf{x}, \mathbf{p}, t) - \sum_a \delta(\mathbf{x} - \mathbf{x}_a(t))\delta(\mathbf{p} - \mathbf{p}_a(t)), \tag{2.3}$$

where the delta-functions tell us that the density is concentrated on the phase trajectories of the individual particles (described by the functions $\mathbf{x}_a(t)$ and $\mathbf{p}_a(t)$, the index a running over all particles from 1 to N). By appropriate integrations of the function f one can "coarse-grain" it, obtaining, e.g., by integration over the momenta, the particle density $n(\mathbf{x}, t)$, the particle current density, etc. By taking various moments of the function f one then obtains a hierarchy of equations (known as the B-B-G-K-Y - or Bogolyubov-Born-Green-Kirkwood-Yvon-hierachy), which yields for each moment an equation whose left-hand side has the form of Eq. (2.3) and the right-hand side contains an integral involving higher moments.

The Boltzmann approximation consists in truncating this hierarchy after the first moment, and replacing the higher moments by a "collision term"; thus, denoting the one-particle distribution function by f_1, the Boltzmann equation becomes

$$\partial_t f_1 + (\mathbf{p}/m)\cdot\partial_{\mathbf{p}} f_1 + q(\mathbf{E} + \mathbf{v} \times \mathbf{B})\cdot\partial_{\mathbf{x}} f_1 = \mathrm{St} f_1 \quad . \tag{2.4}$$

The exact meaning of the density f_1, and the collision term $\mathrm{St} f_1$ (St for "Stoβzahl" - a term introduced by Boltzmann to decribe the number of particles entering or leaving a phase-space cell owing to collisions) depends on a more detailed description. The various other popular approximations in plasma kinetic theory depend on the choice of collision terms, the best known being the Landau collision term and the Balescu-Lenard approximation (cf. a clear description in Lifshitz-Pitayevskii [1979, 1980]).

In what follows I will limit my attention to the Klimontovich equation, but other approximations can be described equally well in this formalism. Thus, the Vlasov-Poisson approximation (cf., e.g., Marsden-Weinstein [1982], Morrison-Weinstein [1982]) consists in replacing f by its first moment and neglecting collisions, as well as assuming that the fields are electrostatic, i.e., $\mathbf{E} = -\operatorname{grad}V$ and $\mathbf{B}$ is a prescribed, external guiding field only. Then the Lorentz-force term simplifies to $-q \operatorname{grad}V.f$ and V satisfies the Poisson equation $\Delta V = -\rho$. I am told by experts that in most practical applications this approximation suffices, and that the fully relativistic formulation discussed below is just a luxury useful perhaps in astrophysical contexts. It should however be remarked that the Klimontovich-Maxwell system is fully Poincare-invariant, fact which is obvious for the Maxwell part, but less so for Eq. (2.4); the invariance can be seen from the fact that a density in momentum space transforms as the fourth (time-) component of a one-form, and hence (2.4) can be written as the vanishing of a Weyl-divergence (codifferential).

To conclude this section let me only note that the applications of plasma physics range from the physics of the ionosphere (on which radio wave propagation and the aurora borealis depend) through the technology which may one day produce controlled nuclear fusion, to the magnetospheres of planets and stars and interstellar ionized gases, important in astrophysics.

3. A POISSON STRUCTURE FOR THE MAXWELL-KLIMONTOVICH SYSTEM

In this section I describe briefly the physicist's version of the Poisson bracket structure for the coupled Maxwell-Klimontovich equations obtained in the thesis of J. C. Hubbard [1982] and described in a paper by I. Bialynicki-Birula and J. C. Hubbard [1981]. I start out with nonrelativistic, 3 + 1 dimensional notation despite the fact that the results are invariant under the Poincaré group.

First we must choose the fundamental dynamical variables. As already mentioned, the prejudice that canonical variables must come in pairs, has to be abandoned for infinite-dimensional "symplectic" manifolds. We choose as the fundamental variables describing the electromagnetic field a pair consisting of one intensity variable $\mathbf{B}(\mathbf{x}, t)$ and one quantity variable $\mathbf{D}(\mathbf{x}, t)$, with the Pauli-Born-Infeld generalized Poisson bracket

$$\{B_i(\mathbf{x}, t), D_j(\mathbf{y}, t)\} = e_{ijk}\partial^k\delta(\mathbf{x} - \mathbf{y}), \tag{3.1}$$

where e_{ijk} is the Levi-Civita tensor, with the gradient of the three-dimensional delta measure in the right-hand side. Note that the times are equal in both fields, and could have been set equal to zero (I will abbreviate the phrase equal-time Poisson bracket by ETPB henceforth). The other ETPB for the electromagnetic field quantities B and D are zero:

$$\{B_i(\mathbf{x}, t), B_j(\mathbf{y}, t)\} = \{D_i(\mathbf{x}, t), D_j(\mathbf{y}, t)\} = 0 . \tag{3.2}$$

The components of the energy-momentum tensor (the generalization of the Hamiltonian to four-dimensional continuum theories), properly defined as the functional derivative of the field Langrangian with respect to the metric tensor (cf. any modern text on relativity, e.g., Landau-Lifshitz, Misner-Thorne-Wheeler, Hawking-Ellis, or Weinberg)

$$T^{\mu\nu}_{EM} = \frac{\delta L}{\delta g_{\mu\nu}} , \tag{3.3}$$

require the introduction of the "other" pair of field quantities: the intensity variable $\mathbf{E}(\mathbf{x}, t)$ and the quantity variable $\mathbf{H}(\mathbf{x}\ t)$. In a nonlinear theory, such as plasma theory, or the Born-Infeld or Mie electrodynamics, the latter are not related to **D** and **B**, respectively, by a linear (albeit tensorial) "constitutive relation", but rather should be considered as "Lagrangian conjugate" of these fields, i.e., the constitutive relations are replaced by the functional derivative definitions:

$$\mathbf{H}(\mathbf{x},t) = \frac{\delta L}{\delta \mathbf{B}(\mathbf{x},t)} , \quad E(\mathbf{x},t) = \frac{\delta L}{\delta \mathbf{D}(\mathbf{x},t)} . \tag{3.4}$$

Note that even though we are in "vacuum" we do not identify **E** with **D**, and **B** with **H** (which is usual when one uses Gaussian units). In three dimensions they are respectively one- and two-forms, associated with circulation and flux; thus, it is easier to think of all electromagnetic quantities as expressed in MKSA units; for simplicity I will put $c = 1$ everywhere.

The electromagnetic energy-momentum-stress tensor is best written in terms of the space-time components (omitting the subscript EM):

$$T^{00} = 1/2[\mathbf{E}\cdot\mathbf{D} + \mathbf{B}\cdot\mathbf{H}] \tag{3.5a}$$

$$T^{0i} = T^{i0} = (\mathbf{E} \times \mathbf{H})^i \tag{3.5b}$$

$$T^{ij} = T^{ji} = 1/2\delta^{ij}[\mathbf{E}\cdot\mathbf{D} + \mathbf{B}\cdot\mathbf{H}] - E^{(i}D^{j)} - B^{(i}H^{j)} , \tag{3.5c}$$

where the parentheses around the upper indices in the last line denote symmetrization, as usual. The explicit use of four field-vectors makes it clear that the energy-momentum quantities are composed of sums of products of intensity and quantity variables, and thus turn out to be densities (in the sense of Weyl), as it should be. We can rewrite $T^{\mu\nu}_{EM}$ in terms of the components of the two-forms F and G, (cf. Thirring [1980] for the vacuum case, where G has been identified with *F), or better still, demonstrate the physical character of this representation by evaluating the appropriate 3-forms (3-dimensional volume, or 2-area × time) on this density to yield the energy density W, momentum density **P** and energy flux (Poynting vector) S.

The energy-momentum tensor of the particles can be expressed in terms of the Klimontovich density (2.4)

$$t^{00} = \sum_a \int d^3pE(\mathbf{p})f_a(\mathbf{x},\mathbf{p},t)$$

$$t^{0j} = \sum_a \int d^3pE(\mathbf{p})p^j f_a(\mathbf{x},\mathbf{p},t) \tag{3.6}$$

$$t^{ij} = \sum_a \int d^3pp^ip^jf_a(\mathbf{x},\mathbf{p},t)$$

This is easily seen to reduce to the usual energy-momentum tensor for a "dust" of particles (cf., e.g., Landau-Lifshitz, or Misner-Thorne-Wheeler):

$$t^{\mu\nu} = (p - e)u^\mu u^\nu - pg^{\mu\nu} \tag{3.7}$$

where p is the pressure, e is the energy-density and u is the four-velocity vector $u^\mu u_\mu = 1$.

Consider now any functionals of the "field variables" f **B**, **D**, which will be denoted by capital letters: F[f, **B**, **D**], G[f, **B**, **D**]. We denote generically any of the components of (f, **B**, **D**) by X(q), with q denoting the seven variables **x**, **p**, t. Generalizing the definition of Poisson bracket from the usual canonical variables (q^i, p_i) to the case of the infinite-dimensional "symplectic" manifold, spanned by the variables (f, **B**, **D**), we have the obvious definition of Poisson bracket for functionals (cf., e.g., Abraham-Marsden [1979], Chernoff-Marsden [1972])

$$\{F,G\} = \sum_{i,j} \frac{\partial F}{\partial X_i} \{X_i, X_j\} \frac{\partial F}{\partial X_j} \quad \text{(discrete case)} \tag{3.8}$$

$$\{F[x],G[x]\} = \int dq\, dq' \frac{\delta F}{\delta X(q)} \{X(q),X(q')\} \frac{\delta G}{\delta x(q')}, \qquad t = t'$$

In terms of the Poisson brackets (1.1) - (1.3) for charged particles in a magnetic field B_{ij}

$$\{p_i, p_j\} = e\, B_{ij}, \text{ etc.} \tag{3.9}$$

which are easily transcribed in terms of the Klimontovich function f, as well as the Pauli-Born-Infeld Poisson brackets (3.1), (3.2) the following set of ETPB is obtained for the variables (f, **B**, **D**):

$$\{f_a(q),f_b(q')\} = \delta_{ab}\big[\partial_{\mathbf{x}}f_a(q)\cdot\partial_{\mathbf{p}}\delta(q-q')-\partial_{\mathbf{p}}f_a(q)\cdot\partial_{\mathbf{x}}\delta(q-q') + $$
$$+ e_a \frac{1}{2} B_{ij}\partial_{p_i} f_a(q)\partial_{p_j}\delta(q-q') \tag{3.10}$$

$$\{f_a(q),\ \mathbf{D}(x,t)\} = -\, e_a \partial_p f_a(q)\delta(x-x') \tag{3.10'}$$

$$\{f_a(q),\ \mathbf{B}(\mathbf{x}',\ t)\} = 0 \tag{3.11}$$

$$\{B^i(\mathbf{x},t),\ D^j(\mathbf{x}',t)\} = e^{ijk}\partial_{x^k}\ \delta(\mathbf{x}-\mathbf{x}') \tag{3.12}$$

$$\{B^i(\mathbf{x},t),\ B^j(\mathbf{x}',t)\} = 0 \tag{3.13}$$

$$\{D^i(\mathbf{x},t),\ D^j(\mathbf{x}',t)\} = 0 \tag{3.14}$$

(where e_a denotes the charge of the species a). From these relations it is easy to see that for arbitrary functionals of the variables (f, **B**, **D**) one can guess the following Poisson bracket, first obtained by Hubbard and Bialynicki-Birula:

$$\begin{aligned}\{F[f,\ \mathbf{B},\ \mathbf{D}],\ G[f,\ \mathbf{B},\ \mathbf{D}]\} &= \int d^3x\ \varepsilon_{ijk}\left(\frac{\partial F}{\partial D_i}\partial_j\frac{\partial G}{\partial B_k} - \frac{\partial G}{\partial D_i}\partial_j\frac{\partial F}{\partial B_k}\right) + \\ &+ \sum_a \int d^6q f_a\Big[\partial_x\frac{\delta F}{\delta f_a}\cdot\partial_p\frac{\delta G}{\delta f_a} - \partial_x\frac{\delta G}{\delta f_a}\cdot\partial_p\frac{\delta F}{\delta f_a} + \\ &+ e_a B_{ij}\frac{\partial}{\partial p_i}\frac{\delta F}{\delta f_a}\frac{\partial}{\partial p_j}\frac{\delta G}{\delta f_a} + e_a\Big(\partial_{p_i}\frac{\delta F}{\delta f_a}\frac{\delta G}{\delta D_j} - \partial_{p_i}\frac{\delta F}{\delta D_j}\frac{\delta G}{\delta f_a}\Big)\Big]\end{aligned} \tag{3.15}$$

This Poisson bracket, derived in this quasi-heuristic manner from the elementary brackets for particles and fields, will be rederived using the momentum map in Section 5. It is easy to show (cf. Hubbard and Bialynicki-Birula [1981]) that these bracket relations and the definition (3.5)-(3.6) of the energy momentum tensor imply that the total energy momentum tensor satisfies the Dirac-Schwinger Poisson bracket relations (note the"Schwinger term" in the right-hand side!):

$$\{T^{00}(\mathbf{x},t),T^{00}(\mathbf{y},t\} = -\left(T^{0k}(\mathbf{x},t) + T^{0k}(\mathbf{y},t)\right)\partial_{x^k}\delta(\mathbf{x}-y) \tag{3.16}$$

$$\{T^{00}(\mathbf{x},t),T^{0k}(\mathbf{y},t)\} = -\left(T^{kj}(\mathbf{x},t) + T^{00}(\mathbf{x},t)\delta^{kj}\right)\partial_{x^k}\delta(\mathbf{x}-\mathbf{y}) \tag{3.17}$$

$$\{T^{0k}(\mathbf{x},t),T^{0j}(\mathbf{y},t)\} = -\left(T^{0j}(\mathbf{x},t)\partial_{x^k} + T^{0k}(\mathbf{x},T)\partial_{x^j}\right)\delta(\mathbf{x}-\mathbf{y})\ . \tag{3.18}$$

From this one can easily verify the explicit Poincare invariance of the theory and get explicit expressions for the momentum, angular momentum, and boost densities for the particle-field system. Since the interaction resides entirely in the Poisson structure it comes as no surprise that the latter are just the sums of the appropriate expressions of the particles and fields.

Finally, in terms of the Hamiltonian density $H = W = T^{00}$, it is easily verified that the canonical evolution equations:

$$\partial_t f = \{f, H\} \tag{3.19}$$

$$\partial_t \mathbf{B} = \{\mathbf{B}, H\} \tag{3.20}$$

$$\partial_t \mathbf{D} = \{\mathbf{D}, H\} \tag{3.21}$$

are identical to the Klimontovich-Maxwell system, provided the equations

$$\mathrm{div}\ \mathbf{B} = 0,\ \mathrm{div}\ \mathbf{D} = \rho = \sum_a e_a \int d^3p f_a(\mathbf{x}, \mathbf{p}, t), \tag{3.22}$$

are treated as constraints, specifying the symplectic manifold.

4. A REVIEW OF MOMENTUM MAPS, APPLIED TO CONTINUA

In this section I review the concept of momentum map (originally introduced by Souriau) and apply it to infinite-dimensional (continuous) dynamical systems. To a certain degree, the material of this section overlaps with other talks at this Conference. It has also been influenced by discussions with Shlomo Sternberg, Daniel Kastler, and Jerrold Marsden, and duplicates some of the results contained in the review by Marsden, Ratiu, Schmid, and Spencer [1982], which I received while I was writing up these notes, and to which the reader is referred for additional details, and a slightly different view. For background material on the momentum map and collective coordinates, cf. Guillemin-Sternberg [1980], and for a physical presentation, from an infinitesimal point of view (albeit in an English which has to be translated back to Russian to be easily understood) Poisson structures for condensed matter are discussed in the paper of Dzyaloshinskii and Volovick [1980] - which came to my attention only after the first draft of these notes was written up.

The momentum map can be viewed as a generalization of Noether's theorem to dynamical systems described by symplectic manifolds or by (finite- or infinite-dimensional) Poisson structures with symmetries. For the case of infinite-dimensional symplectic manifolds treated here, some of the statements are formal, and require a more careful functional-analytic justification, for which we will have to rely on our mathematically more sophisticated colleagues.

I start out with a description of Poisson manifolds which are a generalization of canonical systems (symplectic manifolds). The importance of Poisson brackets in physics has always been emphasized by Dirac, and its role in modern classical mechanics was stressed as in the early 1960s by Jost and Mackey, and more recently the advantage of the Poisson structure approach over the symplectic manifold approach has been emphasized by Hermann [1972, 1973], as well as by the already mentioned authors.

A Poisson manifold P is a (possibly infinite-dimensional) manifold P having a Lie-algebra bracket [f, g] defined on the space of smooth real-valued functions $C^{\infty}(P,R)$. The bracket { , } satisfies the usual requirements for a Lie algebra bracket, i.e., antisymmetry

$$\{f,g\} = - \{g,f\}, \tag{4.1a}$$

the Jacobi identify

$$\{f,\{g,h\}\} + \{g,\{h,f\}\} + \{h,\{f,g\}\} = 0 \ , \tag{4.1b}$$

and is a derivation in each of its arguments, i.e.,

$$\{f,gh\} = h\{f,g\} + g\{f,h\} \ , \tag{4.1c}$$

where gh denotes the usual product of functions. Thus, for functions defined on the usual phase space with canonical coordinates and momenta (q_i, p_i), the Poisson bracket has the form

$$\{f_1,f_2\} = \frac{\partial f_1}{\partial q_i}\frac{\partial f_2}{\partial p_i} - \frac{\partial f_1}{\partial p_i}\frac{\partial f_2}{\partial q_i} \tag{4.2}$$

(summation over repeated indices from 1 to n).

The Poisson bracket may be written in terms of the Lie bracket of vector fields in the following manner: Define the vector field (called the Hamiltonian vector field associated to f)

$$X_f = \frac{\partial f}{\partial p_i}\frac{\partial}{\partial q_i} - \frac{\partial f}{\partial q_i}\frac{\partial}{\partial p_i} \ . \tag{4.3}$$

Then one can write the Poisson bracket of the functions f and g as a Lie bracket of two vector fields:

$$\{f,g\} = -X_f g = +X_g f \text{ and } X_{\{f,g\}} = -[X_f,X_g] \ . \tag{4.4}$$

The minus sign is associated with the "left action" of X_f on g, whereas the "right action" of X_g on f in the second equality goes with a plus sign. This is of no consequence in the "abelian" case under consideration, but plays a role on the case of a general Lie algebra. Although the vector field X_f is uniquely determined by the function f, the converse is not true, since two functions differing by a constant define the same vector field.

The dual space g* of a Lie algebra g, i.e., the space of all (real-valued) linear forms on g*, carries a natural Poisson structure, induced by the Lie bracket [,] of g, also known as the Berezin-Kirillov-Kostant (BKK) Poisson bracket, induced by the coadjoint representation of G. Indeed, let a be an element of g^* and $f: g^* \to R$ be a real-valued function on g^*. We can associate to the derivative of f at a an element of g denoted by f_a (the subscript denotes a partial derivative in the discrete case and a "functional derivative" $f_a = \delta f/\delta a(x)$ in the continuum) such that, with the natural pairing (duality) $\langle\ ,\ \rangle$ between g^* and g we have, for all b in g^*,

$$Df(a)\cdot b = \langle b, f_a\rangle, \tag{4.5}$$

where the left-hand side is the directional derivative of f along the vector b, evaluated at a. For a pair of smooth, real-valued functions F, G on g^* the BKK Poisson bracket is defined by

$$\{F,G\}_-(a) = -\langle a, [F_a, G_a]\rangle , \tag{4.6}$$

where the subscript minus and the minus sign in the right-hand side are associated with the left action of G(i.e., $C^\infty(g^*)$ is identified with the left-invariant functions on the cotangent bundle T^*G of G); in the discrete case the pairing is a sum over the partial derivatives - the usual Poisson bracket, whereas in the continuum case it will be an integral over antisymmetrized products of functional derivatives. For right-invariant functions, the right-hand side in (4.5) carries a + sign, and the bracket so defined is denoted by $\{\ ,\ \}_+$. The notation is flexible enough to handle both the finite-dimensional case and the case where the Lie group G (and the vector spaces g, g^*) may be infinite-dimensional, such as the group of gauge transformations of a gauge theory, or the group of symplectomorphisms (the dynamical group) of a symplectic manifold. In the finite-dimensional case, in terms of a basis (e_i) of g with structure constants c^k_{ij}, where

$$[e_j, e_j] = c^k_{ij} e_k, \tag{4.7a}$$

and the dual basis (e^j), satisfying the bracket relation

$$\{e^i, e^j\} = c^{ij}_k e^k , \tag{4.7b}$$

with $c^{ij}_k = -c^k_{ij}$, Eq. (4.5) becomes:

$$\{F,G\}_-(a) = -c^k_{ij} F_{a_i} G_{a_j} a_k \tag{4.8}$$

where the sum runs over the dimension of g^* and the subscripts denote partial derivatives with respect to the coordinates a_j of the vector a. The bracket with the plus subscript is obtained by changing the sign of the right-hand side, which corresponds to letting the group act on the right on P. In the continuum case the sum is replaced by an integral over the parameters on which a depends (e.g., **x**, **p** in the case discussed in the previous and following sections), and the partial derivatives are replaced by functional derivatives, as in Eq. (3.7), where the role of the structure constants is taken by the brackets of the independent dynamical variables DX(q).

A Hamiltonian (function) on P, is a (smooth, real-valued) function on P giving rise to a Hamiltonian vector field ("Liouville equation") for any f in $C^{\infty}(P)$ describing its "time evolution"

$$\partial_t f = \{f,H\} = - X_H f \quad . \tag{4.9}$$

One may view X_H as the vector field on P associated to the covector field dH by the natural Poisson pairing defined by the fundamental two-form of the dynamical system (cf., e.g., Abraham-Marsden [1978], Arnol'd ([1978], or Souriau [1970]).

We now go on to discuss the symmetries of Poisson mnifolds. Let the Lie group G act on the Poisson manifold P smoothly on the left (or right) preserving the Poisson brackets, i.e., if we denote by $s:P \to P$ the action of G on P, the induced action on s*F on a function F in $C^{\infty}(P)$ must satisfy

$$\{s^*F, s^*G\} = s^*\{F, G\} \tag{4.10}$$

for all F, G in $C^{\infty}(P)$. I remind the reader that a group action on a symplectic manifold is symplectomorphism if it preserves the symplectic structure, i.e., denoting the two-form defining the symplectic structure on M by ω:

$$s^*\omega = \omega \tag{4.11}$$

In other words, the group action is a "canonical transformation" (in the sense used in the physics literature). The adjoint action of the Lie group G on the Lie algebra g is denoted by Ad(g): b Ad(g)b. The transpose of this action, more precisely, the action on an element a in g* defined by the pairing of g* and g, for arbitrary b in g:

$$\langle Ad^*(g)a, b\rangle = \langle a, Ad(g^{-1})b\rangle \tag{4.12}$$

is called the coadjoint representation of G on g*. A Hamiltonian action (also called Poisson action) of G on a Poisson algebra P is a linear representation $r:G \to GL(P)$ together with a linear map $a \to f(a)$ of g* into P with the properties (Guillemin-Sternberg [1980], Marsden, Ratiu, Schmid, and Spencer [1982])

$$r(g)(f(a_1)f(a_2)) = (r(g)f(a_1))\cdot(r(g)f(a_2)) \tag{4.13}$$

$$r(g)\{f(a_1), f(a_2)\} = \{r(g)f(a_1), r(g)f(a_2)\} \tag{4.14}$$

$$f([a_1,a_2]) = \{f(a_1), f(a_2)\} \tag{4.15}$$

$$r(g)f(a) = f(Ad(g)a) \quad . \tag{4.16}$$

The coadjoint representation of G on g* produces a Hamiltonian action on $C^\infty(g^*)$. For an arbitrary Poisson structure $(P, \{\ ,\ \})$ with a Hamiltonian action of G on P we can construct a mapping $S: P \to g^*$ of the manifold into the dual of the Lie algebra, called the (Souriau) momentum map (some authors use the gallicism "moment map") which for a point p of P associates to each element a of the Lie algebra g the value of the Hamiltonian function f(a) at the point p:

$$\langle S(p), a \rangle = f(a)(p) \tag{4.17}$$

Since this function is linear in a, it is an element of g* associated to the point p of P. The name "momentum map" comes from the fact that if G is the translation group of R^3, both g and g* can be identified with R^3 and a point $(\mathbf{q}, \mathbf{p})$ of the phase space of a particle is mapped into the 3-momentum vector $\mathbf{p}$ of the particle. For the rotation group the momentum map yields the angular momentum, and thus the existence of the momentum map is a generalization of Noether's theorem to Poisson actions of groups on Poisson manifolds which do not necessarily stem from an invariant action principle, so there is no Noether's theorem.

The momentum map has a number of obvious properties, such as equivariance, smoothness (if the Hamiltonian action is smooth), and posesses an important extension to semidirect products, which are important for applications. In the case of fluids, plasmas, and other continua, the Poisson manifold is infinite-dimensional, the group G is a subgroup of the group of diffeomorphisms of the manifold, functions become "functionals" - hence the functional derivative notation has to be used. We now pass to the case under consideration and rederive the Klimontovich-Maxwell Poisson structure by means of the momentum map in the following section. For other applications of the momentum map and Poisson structures to continua the reader is referred to the articles by Dzyaloshinskii and Volovick [1980] (condensed matter physics), Guillemin and Sternberg [1980] (collective coordinates, nuclear physics, semidirect products, general discussion), Marsden, Ratiu, Schmid, and Spencer [1982], Marsden, Ratiu and Weinstein [1982], (fluids, Vlasov-Poisson plasmas, general theory), Mayer [1983] (relativistic superfluids), Weinstein [1981-1982] (general theory, gauge field theory), and several other papers in this volume. Additional references can be found in these papers.

5. THE MOMENTUM MAP AND POISSON STRUCTURE FOR PLASMAS

For the dynamics of continua the symmetry group G is an infinite-dimensional subgroup of the dynamical group of the system, elements of the dual space should be interpreted as distributions, or measures. Thus, the momentum map itself, a smooth mapping from P to g* will be represented by either a distribution or by a measure (an integral) and therefore we expect the Poisson structure to involve integrals. Consider first the simple case where the independent elements of g* are (smooth) functions X(q) on a phase space Q, satisfying the (postulated) bracket relations

$$\{X(q'), X(q'')\} = \int c(q,q',q'')X(q)dq \quad . \tag{5.1}$$

In order to determine the "structure constants" c(q, q', q"), one should proceed as in the finite-dimensional case, and first determine the "structure constants "C(q, q', q") (which differ from c(q, q', q") only by a sign) of the infinite-dimensional Lie algebra of the symmetry group G. As far as Maxwell's equations are concerned, the symmetry group is the infinite-dimensional group of gauge transformations of classical electrodynamics, whereas the Klimontovich function (or any other Liouville density in an appropriate relativistic phase space) the group is the group of canonical transformations. When these two groups are combined and potentials are used, the overall group is a semidirect product, since the gauge transformations affect the canonical momenta in the presence of electromagnetic potentials. This fact was exploited by Marsden and Weinstein [1982] in their derivation of the Poisson structure for the Maxwell-Vlasov system. However, since they made use of potentials which are subject to gauge-fixing constraints, they had to utilize the Marsden-Weinstein reduction technique (Marsden and Weinstein [1974]), and subsequently eliminate the potentials. The Poisson structure derived heuristically in Section 3 could also have been obtained via the quantum theory approach advocated by Dzyaloshinskii and Volovick [1980] -- in fact an attempt to do this, by replacing the Klimontovich function f by the Wigner distribution function on quantum-mechanical phase space and using the Wigner-Moyal brackets, was undertaken in 1979 by Hubbard and the author; it was not completed because of our failure to recognize the usefulness of the Pauli-Born-Infeld commutation relations (3.1)-(3.2) for the quantized electromagnetic field strengths, to which we were converted only in 1981 by Iwo Bialynicki-Birula.

Let us try to construct the symmetry group of the dynamical system described by the functions $f(\mathbf{x}, \mathbf{p}, t)$, $\mathbf{B}(\mathbf{x}, t)$, and $\mathbf{D}(\mathbf{x}, t)$ without appealing directly to potentials and gauge transformations. The space on which the relativistic theory unfolds is the 7-dimensional (or 7n-dimensional) phase space, obtained from the 8-dimensional space of four-coordinates and four-momenta by restricting the momenta to lie on the mass shells of the particles (Kovacich [1977], Kovacich and Mayer [1978]). The first apparent difficulty is that the dynamical variables **B** and **D** depend only on the 4 coordinates, but not on momenta. This is not a real difficulty, since one may think of these variables defined on "cylinders" in the 8n-dimensional space, constant on the momentum variables. Since the elements of the Lie group and the Lie algebra are smooth functions on these spaces, assumed to have certain decay properties at infinity, the elements of the dual space will be distributions or measures (in the case of the Klimontovich distribution function it should be noted that the six-form fd^3xd^3p is a measure (or the physicist's representation of a distribution - more correctly, a de Rham current), hence its dual elements f* are continuous (or smooth) functions of appropriate growth at infinity; similarly the electromagnetic field strengths should be viewed as two-forms, so that their duals may be considered bivector densities (or de Rham currents) integrated over the four-dimensional space).

For pedagogical reasons it is convenient to discuss first the Poisson structure for the Klimontovich density f in the absence of the electromagnetic field, which will illustrate the measure-theoretic or distributional aspects involved, at a manageable level, since there are no indices, two-forms, or bivectors involved. I then discuss the Pauli-Born-Infeld Poisson structure for the free electromagnetic field, and finally, the mutual modification of the Poisson brackets by coupling the two fields together. As already mentioned, the energy-momentum tensor for the coupled system is just the sum of the field and particle tensors, since the whole interaction resides in the Poisson structure. The price paid for proceeding in a gauge-independent way, without using canonical momenta and potentials is minimal, if one is willing to enlarge the framework and consider Poisson brackets which have differentials of delta-currents in the right-hand side. This is meaningful, if one considers the various field-quantities as de Rham currents, defined over appropriate test-forms. It should also be useful in a discussion of quantum field theories on manifolds, to which I will return elsewhere. The discussion of the Poisson structure for f is similar to the one given by Marsden and Weinstein [1982].

For simplicity, we consider a single species of particles, and by setting their mass masses equal to one, there is no need to distinguish between velocity and momentum. The 6-dimensional phase space has the canonical symplectic structure defined by the two-form $\omega = dx^1 \Lambda\ dp_i$. The symmetry group of the system is the group S of all symplectomorphisms (or canonical transformations), with restrictions imposed on the growth at infinity of the generating functions. I assume that the elements of S, or more precisely, the generating functions which are the elements of the Lie algebra have moderate growth both in coordinates and momenta, so that the elements of the dual space are rapidly decreasing distributions on phase space. Obviously the appropriate pairings are finite - the less restrictive one is in one space the more restrictive one will have to be in the dual space. The Lie algebra s will consist of the generating functions of canonical transformations (forgetting the fact that these are defined up to affine transformations, i. e., additive constants and scaling) with the mentioned growth restrictions, with the Lie algebra structure given by the classical Poisson brackets on phase space, Eq. (4.2)

$$[f,g] = - \{f,g\} \tag{5.2}$$

and the negative sign agreeing with our previous convention about group actions. The dual space s* consists of distributions on s with sufficiantly rapid decrease, so that it is legitimate to represent them by density functions, as is usual in the physics literature

$$\langle f,G\rangle = \int f(q)G(q)d^3x\, d^3p \tag{5.3}$$

(the six-form $G = G(q)d^3x \wedge d^3p$ is the de Rham current dual to f(q)). In order to get the BKK bracket for the densities f(q), G(q), we should use Eq. (4.5), and for this it is necessary to write the action of the Hamiltonian vector field of f on the current G (cf. Eq. (4.3) for the definition of X_f):

$$X_f G = - \{f, \frac{\delta G}{\delta f}\} \quad . \tag{5.4}$$

This is essentially an "integration by parts" argument, which together with the Berezin-Arnol'd-Kirillov-Kostant-Souriau construction of the bracket leads to the Poisson structure for functionals F, G of the Klimontovich function f

$$\{F,G\} = \int f(q) \{\frac{\delta F}{\delta f}, \frac{\delta G}{\delta f}\}(q)\, d^3x d^3p \quad . \tag{5.5}$$

It is easy to see that this is only a fancy way of writing the canonical Poisson brackets between the coordinates and momenta if one replaces F and G by the Klimontovich expression and evaluates the delta functions. On the other hand, the bracket survives the transition to other distribution functions, such as the Boltzmann, or Vlasov distribution. It obviously satisfies the Jacobi identity, and its interpretation in terms of coadjoint orbits can be found in Marsden and Weinstein [1982].

We now recast the Poisson brackets of the electromagnetic field without sources into the same form. The simplest approach is to use the four-potential one-form A, such that F = dA. But this one-form is subject to gauge-transformations, i.e.,

$$A \longmapsto A + dy \tag{5.6}$$

where y is an arbitrary smooth function. One subjects A to the Lorentz gauge condition $d^*A = 0$, which imposes on y the condition that it be a solution of the wave equation $d^*dy = 0$. This still leaves some freedom, and one may proceed in a fixed gauge $A_0 = 0$, div**A** = 0 (transverse gauge), and choose **A** and its canonical conjugate D as the canonical variables (the second pair of Maxwell equations d*G = 0 appears as a "Bianchi identity"). Proceeding in this manner and using the group of gauge transformations and its momentum map, combined with the Marsden-Weinstein [1974] reduction technique and integration by parts, leads to the Poisson bracket for functionals F and G of **B** and **D** (recall that $\mathbf{B} = \mathrm{curl}\mathbf{A}$)

$$\{F[\mathbf{B},\mathbf{D}],\ G[\mathbf{B},\mathbf{D}]\} = \int \left(\frac{\delta F}{\delta \mathbf{D}} \cdot \nabla \times \frac{\delta G}{\delta \mathbf{B}} - \frac{\delta G}{\delta \mathbf{D}} \cdot \nabla \times \frac{\delta F}{\delta \mathbf{B}}\right) dq \quad , \qquad (5.7)$$

i.e., the last term of Eq. (3.16). Substituting **B** and **D** for the functionals F and G we are led to the Pauli-Born-Infeld Poisson brackets (3.1), (3.2), which were the starting point of Section 3.

In order to derive the "cross-terms" of Eqs. (3.9) - (3.16) one may again introduce the vector potential **A**, remembering that the canonical momenta (in distinction from the kinetic momenta used so far) are functions of **A**: $\mathbf{p}_{can} = \mathbf{p} - e\mathbf{A}$. Eliminating the potentials and making use of the gauge-variance of the canonical momenta, one is naturally led to the remainder of the terms in (3.9)-(3.16).

However, this derivation contradicts the spirit of the preceding sections, and therefore I will describe in words (since a detailed discussion would take up too much space) how one could obtain the complete Poisson structure (3.9) - (3.16) directly in terms of **B** and **D**, without the detour to potentials. For this we postulate the Born-Infeld relations (3.1) - (3.2) as well as the bracket (3.8) describing the particle-field interaction, and then proceed just as we did for the Klimontovich function f, by first defining a Lie bracket (in the Lie algebra of infinite-dimensional vector fields on the manifold under consideration), and then dualizing it by means of the BKK relation (4.6). Since the quantities involved are smooth functions with definite growth or decay properties at infinity, it is natural that the dual objects will be measures or distributions, defined in such a manner that the integrals converge. A rather lengthy calculation then show that the correct brackets are indeed the ones given in Section 3. For an example of such a derivation I refer the reader to my paper Mayer [1983] where such an approach is used.

6. CONCLUSIONS

In addition to the concrete benefits that can be derived from the Poisson structure for the relativistic Klimontovich-Maxwell system, such as the possibility of deriving a more rapidly convergent perturbation theory for laboratory plasmas, or the treatment of special-relativistic or even general-relativistic astrophysical plasmas one can draw some general conclusions, not specifically related to the system under consideration.

1. It is not only convenient, but in some cases imperative, to treat interactions, particularly with gauge fields, by means of a modification of the underlying symplectic structure, or Poisson structure in the classical theory. Not only does this simplify the treatment (the Hamiltonian, or its counterpart the stress-energy-

momentum tensor, decomposes into a sum of free-field or free-particle Hamiltonians, and thus it becomes particularly easy to write the equations of motion and verify their invariances).

2. The quantum-theoretic analog of this statement can be roughly formulated in the following manner: instead of introducing the interaction between various quantized fields by an interaction Lagrangian or Hamiltonian, the interaction is contained in a modification of the canonical commutation relations of the theory. This becomes particularly transparent in the treatment of particle motion in prescribed magnetic fields, such as the treatment of the quantum Hall effect, which has attracted considerable attention lately. In a full-fledged quantum theory the field which is modifying the CCR or CAR is itself a quantized field, but this approach still remains to be developed. It could apply, in particular, to the interaction of matter with Yang-Mills fields. This might lead to an "algebraic" approach to quantum gauge theories.

3. For practical applications it might be useful to introduce something akin to the Dirac (interaction) picture: the "large" part of the interaction is included in the Poissn structure (or the commutation relations), and assuming this problem can be solved, small deviations from this "guiding field interaction" are then treated as perturbations to the Hamiltonians describing the evolution of the system.

At any rate, it is important that physicists become familiar with these new differential-geometric approaches, since they seem not only to simplify computations, but also shed new light on old theories. They might be particularly useful in theories which have no linear (or free-field) approximation.

ACKNOWLEDGMENTS

I would like to express my gratitude to the following colleagues, discussions with whom over the past few years have influenced my views on the subject: Iwo Bialynicki-Birula, John C. Hubbard, Allan Kaufman, Daniel Kastler, Michael Kovacich, Jerrold Marsden, and Shlomo Sternberg. To the latter I would also like to express my appreciation for the superb organization of the Conference, and for his warm hospitality in Jerusalem.

REFERENCES

Abraham, R., and J. Marsden [1978], Foundations of Mechanics, second edition, Addison-Wesley, Reading MA.

Arnol'd, V. I. [1974, 1978], Matematichskie metody klassicheskoi mekhaniki [Mathematical Methods of Classical Mechanics], Nauka, Moscow, 1974; English translation: Springer Verlag New York-Berlin-Heidelberg, 1978. Appendix 5.

Berezin, F. [1967], Some remarks about the associated envelope of a Lie algebra, Funct. Anal. Appl. **1**, 91-102.

Bialynicki-Birula, I., and J. C. Hubbard [1981] Gauge-independent canonical formulation of relativistic plasma theory, U. C. Irvine Preprint and Phys. Rev. A to be published.

Born, M. and L. Infeld [1935], On the quantization of the new field theory, Proc. Roy. Soc. (Lond.) **A 150**, 141 (and references there to classical nonlinear electrodynamics of 1934); cf also Pauli [1932].

Dzyaloshinskii, I. V. and G. E. Volovick [1980], Poisson brackets in condensed matter physics, Ann. Phys. (NY) **125**, 67-97.

Chernoff, P., and J. Marsden [1974], Properties of infinite-dimensional Hamiltonian Systems, Lecture Notes in Mathematics Vol. 425, Springer Verlag, Berlin-Heidelberg-New York, 1978.

Guillemin, V., and S. Sternberg [1980], The moment map and collective motion, Ann. Phys. (NY) **127**, 220-253.

Hermann, R. [1972, 1973], J. Math. Phys. **13**, 833-878; Differential geometric methods in physics and engineering, Interdisciplinary Mathematics, Math. Sci. Press, Brookline, MA, 1973, and many subsequent volumes of the Series.

Hubbard, J. C. [1982], Gauge-independent Hamiltonian formulation of relativistic plasma theory, Ph. D. Thesis, U. C. Irvine, Unpublished.

Hubbard, J. C., and M. E. Mayer [1981], Kinetic theory on the tangent bundle, Bull. Amer. Phys. Soc. **26**, 855.

Kaufman, A. N., and R. G. Spencer [1982], Hamiltonian structure of two-fluid plasma dynamics, Phys. Rev. **A 25**, 2437-2439.

Kirillov, A. A. [1972, 1976], Unitarnye predstavleniva nil'potentnykh grupp Lie (Unitary representations of nilpotent Lie groups), Uspekhi Mat. Nauk 1962, **17**(4), 57-101 [Transl. Russian Math. Surveys]. Elementy teorii predstavlenii (Elements of representation theory), "Nauka", Moscow, 1972 [Transl: Springer Verlag, 1976].

Klimontovich, Yu. L. [1975] Kineticheskaya teoriva neideal'nogo gaza i neideal'noi plazmy (Kinetic theory of a nonideal gas and a nonideal plasma) "Nauka", Moscow: [There exists an English translation].

Kostant, B., in Lecture Notes in Mathematics Vol. 170, Springer Verlag Berlin Heidelberg New York, 1970.

Kovacich, M. [1977], Ph. D. Thesis, U. C. Irvine, Unpublished.

Kovacich, M., and M. E. Mayer [1978], Relativistic Statistical Mechanics: BBGKY Hierarchy and Boltzmann Equation, Ann. Israel Phys. Soc. **2**, 928-931.

Lifshitz, E. M., and L. P. Pitaevskii [1979] Fizicheskaya kinetika (Physical kinetics), [vol. X of Landau-Lifshitz], "Nauka", Moscow [English translation: Pergamon, London-New York, [1980].

Littlejohn, R. G. [1981], J. Math. Phys. **20**, 2445; Phys. Fluids **24**, 1730. U. C. Berkeley Thesis, 1978.

Marsden, J. [1982], A group-theoretic approach to the equations of plasma physics, Canad. Math. Bull.**99**, 1 - 14.

Marsden, J., T. Ratiu, R. Schmid, and R. Spencer [1982], Hamiltonian systems with symmetry, coadjoint orbits, and plasma physics, UC Berkeley Preprint, 1982; to appear in Proceedings of the IUTAM Symposium on Modern Developments in Analytical Mechanics, Italy, June 7-11, 1982.

Marsden, J., T. Ratiu, and A. Weinstein [1982], Semi-direct Products and Reduction in Mechanics, Berkeley Preprint PAM-96, September 1982.

Marsden, J., and A. Weinstein Reduction of symplectic manifolds, Rep. Math. Phys. (Torun), 1974, **5**, 121-130; The Hamiltonian structure of the Maxwell-Vlasov equations, Physica 1982, **D4**, 394-406; Coadjoint orbits, vortices, and Clebsch variables for incompressible fluids, Physica **D**, 1982, to appear.

Mayer, M. E. [1983], A Poisson structure for relativistic superfluids, Preprint, U. C. Irvine Phys. Dept., March 1983, submitted for publication.

Morrison, P. J. [1980], The Maxwell-Vlasov equations as a continuous Hamiltoniam system, Phys. Lett. **80A**, 383-386.

Morrison, P. J., and A. Weinstein [1981],Comments on: Poisson brackets for the Maxwell-Vlasov equations Phys. Lett. **86A**, 235-236.

Novikov, S. P. [1982], Gamil'tonov formalism i mnogoznachnyi analog teorii Morsa (Hamiltonian formalism and the multivalued analog of Morse theory) Usp. Matem. Nauk **37** (5), 3 -49 [Engl. Transl: Russian Math. Surveys, to appear, 1983].

Pauli, W. [1932, 1981], Die allgemeinen Prinzipien der Quantenmechanik in Handbuch der Physik, vol. 24, Springer Berlin. English Translation: The general principles of quantum mechanics, Springer Verlag, New-York, Berlin, Heidelberg, 1981.

Souriau, J. M. [1970], La structure des systèmes dynamiques, Dunod, Paris, 1970.

Sternberg, S. [1978], in "Differential-Geometric Methods in Physics", edited by K. Bleuler and A. Reetz, Lecture Notes in Mathematics No. 676, Springer Verlag, Berlin/Heidelberg/New York, 1978.

Thirring, W. [1980]; Lehrbuch der Mathematischen Physik, Vol. II Springer, Vienna; English Translation, Springer, New York, 1982.

Weinstein, A. [1981, 1982], Symplectic geometry, Bull. Amer. Math. Soc. **5**, 1-13. Gauge groups and Poisson brackets for interacting particles and fields, in "Mathematical Methods in Hydrodynamics...", M. Tabor and Y. Treve, eds. AIP Conf. Proceedings **88**, 1-11.

NOETHER'S THEOREM FOR HARMONIC MAPS

John Rawnsley

Mathematical Institute
University of Warwick
Coventry CV4 7AL, England

We determine the 1-form which is conserved as a consequence of Noether's theorem for symmetries of harmonic maps. When the symmetries arise from isometries of the codomain this 1-form is the pull-back of the momentum map on the tangent bundle. If the codomain is homogeneous the conservation of this 1-form is not only necessary but also sufficient.

1. INTRODUCTION

Harmonic maps [1] are simultaneously generalizations of harmonic functions on and geodesics in a Riemannian manifold. If (M,g) and (N,h) are Riemannian manifolds with $\varphi : M \longrightarrow N$ a smooth map, then the energy of φ is defined to be

$$E(\varphi) = \tfrac{1}{2}\int_M |d\varphi|^2 \, v_g$$

where $d\varphi$ is the differential of φ viewed as a 1-form on M with values in the pull-back $\varphi^{-1}TN$ of the tangent bundle of N (that is: as a section of $T^*M \otimes \varphi^{-1}TN$) and $|d\varphi|^2$ is calculated using the dual of g on T*M and the pull-back $\varphi^{-1}h$ on $\varphi^{-1}TN$. φ is harmonic if it is a critical point of this functional. We shall determine the Euler-Lagrange equations which φ must satisfy below.

Taking M to be $\mathbb{R}$ or S^1 with their standard metrics, a harmonic map is a geodesic in N, whilst taking N to be $\mathbb{R}$ a harmonic map is a harmonic function. Minimal sur-

S. Sternberg (ed.), Differential Geometric Methods in Mathematical Physics, 197–202.

faces are also examples of harmonic maps with two-dimensional domain, as are the non-linear σ-models which have recently been studied by physicists as examples of non-linear field theories.

Hamiltonian methods have long been used in the study of geodesics so it is natural to ask if symplectic geometry has anything to offer in the case of higher dimensional domain. At the time of this lecture I know of no such general application, but I shall determine the conserved current given by Noether's theorem when the harmonic map problem admits symmetries and interpret this in terms of a momentum map. I cannot yet exploit this to yield anything new.

2. THE CONSERVED 1-FORM

Noether's theorem in the Calculus of Variations comes from considering variations of an extremal which keep the Lagrangian density constant, or, more generally, variations which can be compensated for by means of a transformation in the domain. See [2] for details in the general case.

Let us carry out the calculations in detail in the case of harmonic maps. Thus (M,g) and (N,h) are Riemannian manifolds and $\varphi : M \longrightarrow N$ a smooth map. The energy of φ is given by

$$E(\varphi) = \int_M e(\varphi)\, v_g$$

where v_g is the Riemannian volume and $e(\varphi)$ the energy density

$$e(\varphi) = \tfrac{1}{2}|d\varphi|^2$$

where $d\varphi$ is viewed as a section of $T^*M \otimes \varphi^{-1}TN$.

Let φ_t be a 1-parameter variation of φ, so $\varphi_o = \varphi$, and let σ_t be a 1-parameter family of diffeomorphisms of M. Put

$$\nu = \dot{\varphi}_o \in C^\infty(\varphi^{-1}TN), \quad \xi = \dot{\sigma}_o \circ \sigma_o^{-1} \in C^\infty(TM)$$

then for any exterior form α on M

$$\sigma_o^*(\mathcal{L}_\xi \alpha) = d/dt_{\,t=o}\, \sigma_t^* \alpha .$$

Thus

$$d/dt\big|_{t=o}\ \sigma_t^*\{e(\varphi_t)v_g\} = \sigma_o^*\mathcal{L}_\xi(e(\varphi)v_g) + \sigma_o^* d/dt\big|_{t=o} e(\varphi_t)v_g.$$

By the results in [1],

$$d/dt\big|_{t=o} e(\varphi_t) = \langle d\nu, d\varphi\rangle$$

where $d\nu$ is the covariant exterior differential of ν. But

$$\langle d\nu, d\varphi\rangle = -d^*[(\nu \lrcorner \varphi^{-1}h)\circ d\varphi] - (\varphi^{-1}h)(\nu, \tau_\varphi)$$

where $\tau_\varphi = -\nabla^* d\varphi$ is the tension field of φ, and so

$$d/dt\big|_{t=o}\ \sigma_t^*(e(\varphi_t)v_g) = \sigma_o^*\{d[e(\varphi)\xi\lrcorner v_g - d^*[(\nu\lrcorner \varphi^{-1}h)\circ d\varphi]v_g - (\varphi^{-1}h)(\nu, \tau_\varphi)v_g\}.$$

If we take $\sigma_t \equiv id_M$ we recover the formula for the first variation of the energy

$$d/dt\big|_{t=o} E(\varphi_t) = -\int_M (\varphi^{-1}h)(\nu, \tau_\varphi)v_g$$

which vanishes for all variations if and only if

$$\tau_\varphi = 0$$

which is the harmonic map equation.

Returning to the general case,

$$\xi\lrcorner v_g = *\xi\lrcorner g$$

so

$$\begin{aligned} d[e(\varphi)\xi\lrcorner v_g] &= d*[e(\varphi)\xi\lrcorner g] \\ &= **d*[e(\varphi)\xi\lrcorner g] \\ &= -*d^*[e(\varphi)\xi\lrcorner g] \\ &= -d^*[e(\varphi)\xi\lrcorner g]v_g. \end{aligned}$$

Hence

$$d/dt\big|_{t=o}\ \sigma_t^*(e(\varphi_t)v_g) = -\sigma_o^*\{[d^*(e(\varphi)\xi\lrcorner g + (\nu\lrcorner \varphi^{-1}h)\circ d\varphi) + (\varphi^{-1}h)(\nu, \tau_\varphi)]v_g\}.$$

To obtain Noether's theorem we suppose φ is harmonic so $\tau_\varphi = 0$, and $\sigma_t{}^*(e(\varphi_t)v_g)$ is independent of t, then

$$d^*\{e(\varphi)\xi \lrcorner g + (\nu \lrcorner \varphi^{-1}h)\circ d\varphi\} = 0.$$

Hence we have the

THEOREM. Let $\varphi : M \to N$ be harmonic, φ_t a variation of φ and $\sigma_t : M \to M$ a 1-parameter family of diffeomorphisms such that $\sigma_t{}^*(e(\varphi_t)v_g)$ is constant then the 1-form on M

$$e(\varphi)\xi \lrcorner g + (\nu \lrcorner \varphi^{-1}h)\circ d\varphi$$

is coclosed where $\xi = \dot\sigma_o \circ \sigma_o{}^{-1}$, $\nu = \dot\varphi_o$.

If we do not assume immediately that φ is harmonic but that $\sigma_t{}^*(e(\varphi_t)v_g)$ is constant, then we get the formula

$$(1) \qquad \varphi^{-1}h(\nu, \tau_\varphi) = -d^*\{e(\varphi)\xi \lrcorner g + (\nu \lrcorner \varphi^{-1}h)\circ d\varphi\}.$$

This can be used to check in particular cases if the necessary condition in the theorem is also sufficient.

3. ISOMETRIES OF THE CODOMAIN

Suppose N admits a 1-parameter group ρ_t of isometries, then $\varphi_t = \rho_t \circ \varphi$ is a variation with $e(\varphi_t) = e(\varphi)$ for all t and $\nu = \varphi^{-1}Y$ where Y is the infinitesimal generator of ρ_t. Taking $\sigma_t = \mathrm{id}$ for all t then $\xi = 0$ so (1) gives

$$(2) \qquad (\varphi^{-1}h)(\varphi^{-1}Y, \tau_\varphi) = -d^*\{\varphi^*(Y \lrcorner h)\}$$

so φ harmonic implies

$$(3) \qquad d^*(\varphi^*(Y \lrcorner h)) = 0.$$

Conversely, if (3) holds then (2) implies τ_φ is orthogonal to the Killing field Y. Hence if N is a homogeneous Riemannian manifold then $\varphi : M \to N$ is harmonic if and only if (3) holds for all Killing vector fields Y.

If N admits the Lie group G of isometries with Lie algebra $\mathfrak{g}$, and for each $\xi \in \mathfrak{g}$, $\tilde\xi$ denotes the corresponding Killing vector field defined by

$$\tilde{\xi}_y = d/dt\big|_{t=o} \exp -t\,\xi . y$$

then we may define a 1-form μ on N with values in $\mathfrak{g}^*$ the dual vector space of $\mathfrak{g}$, by

$$\langle\mu(X), \xi\rangle = -h(X, \tilde{\xi}), \qquad X \in \mathfrak{X}(N),\ \xi \in \mathfrak{g}.$$

Then (2) may be rewritten

$$(4) \qquad \varphi^{-1}\mu(\tau_\varphi) = -d^*(\varphi^*\mu).$$

μ is well-known in symplectic geometry as the momentum map

$$\mu : TN \longrightarrow \mathfrak{g}^*$$

for the symplectic action of G on TN, see [3].

Suppose N is a G-invariant submanifold of an orthogonal representation of G on a vector space V and the embedding of N in V is isometric. Suppose $\mathfrak{g}$ carries an invariant metric (,), then there is a map

$$VxV \longrightarrow \mathfrak{g}$$

denoted by $v_x w$ for v,w in V defined by

$$(v_x w, \xi) = (\xi . v, w)_V, \qquad \xi \in \mathfrak{g}, \quad v, w \in V$$

where $\xi . v$ denotes the infinitesimal action of $\mathfrak{g}$ on V. Then for $x \in N$, $X \in T_xN \subset V$,

$$\begin{aligned}\langle\mu(X), \xi\rangle &= -h(X, \tilde{\xi}_x)\\ &= h(X, \xi . x)\\ &= (X, \xi . x)_V\\ &= (x_x X, \xi).\end{aligned}$$

Thus if we identify $\mathfrak{g}$ with $\mathfrak{g}^*$ via (,) and view μ as a map from TN to $\mathfrak{g}$ then

$$\mu(X) = x_x X \qquad \text{if} \quad X \in T_xN.$$

Thus from (4) we have

$$\begin{aligned}-\varphi^{-1}\mu(\tau_\varphi) &= d^*(\Phi_x d\Phi)\\ &= -\sum_i X_i \lrcorner \nabla_{X_i}(\Phi_x d\Phi)\end{aligned}$$

$$= - \sum_i X_i \lrcorner (X_i(\Phi) \times d\Phi + \Phi \times \nabla_{X_i} d\Phi)$$

$$= - \sum_i X_i(\Phi) \times X_i(\Phi) + \Phi \times X_i \lrcorner \nabla_{X_i} d\Phi ,$$

where Φ is the composition of φ with the inclusion of N in V and $X_1, \ldots, X_m$ is a local orthonormal frame field on M. But $v \times w$ is skew symmetric so the first term in the summation is zero, giving

$$- \varphi^{-1}\mu(\tau_\varphi) = \Phi \times \Delta\Phi$$

with $\Delta\Phi$ calculated as a vector-valued function on M. Hence φ harmonic implies

$$\Phi \times \Delta\Phi = 0$$

with the converse holding if G acts transitively on N.

As a particular case take $G = O(n)$, $V = \mathbb{R}^n$, $N = S^{n-1}$ then $v \times w \in o(n)$ is the skew symmetric matrix

$$(v \times w)_{ij} = v_i w_j - v_j w_i$$

so

$$v \times w = 0 \quad \Leftrightarrow \quad v, w \text{ linearly dependent},$$

hence

$$\Phi \times \Delta\Phi = 0 \quad \Leftrightarrow \quad \Delta\Phi \,//\, \Phi$$

so we recover the well-known result that a map $\varphi : M \longrightarrow S^{n-1}$ is harmonic if and only if the function $\Phi : M \longrightarrow \mathbb{R}^n$ satisfies

$$\Delta\Phi = \lambda\Phi$$

for some function λ on M, which is easily found to be $2e(\varphi)$.

REFERENCES

1. Eells,J. and Lemaire,L. "Selected Topics in Harmonic Maps". To appear in the CBMS Regional Conference Series.

2. Goldschmidt,H. and Sternberg,S. "The Hamilton-Cartan Formalism in the Calculus of Variations". Ann. Inst. Fourier Grenoble 23 (1973) 203-267.

3. Guillemin,V. and Sternberg,S. "Geometric Asymptotics". A.M.S. Mathematical Surveys number 14, Providence 1977.

WAVE FUNCTIONS AND TRANSVERSE MEASURES

D.J. Simms

School of Mathematics, Trinity College, Dublin

The inner product of wave functions in geometric quantisation using transverse measures is considered, as a way of admitting polarisations with non-reducible foliations.

INTRODUCTION

One of the fundamental requirements for a theory of geometric quantisation is a Hilbert space structure on the space of wave functions. The purpose of this note is to consider the use of transverse measures in the sense of Connes (1) as a way of allowing integration over singular spaces.

Let M be a symplectic manifold of finite dimension 2n with symplectic form ω, representing a classical phase space. Let T be the tangent bundle of M and $T^{\mathbb{C}}$ and its complexification. The construction of the wave functions depends on the choice of a polarisation F of (M,ω). Thus F is an involutive sub-bundle of $T^{\mathbb{C}}$ of constant fibre dimension n and isotropic with respect

S. Sternberg (ed.), Differential Geometric Methods in Mathematical Physics, 203–206.

to ω. We also require that the real sub-bundle $D = F \cap \bar{F} \cap T$ have constant fibre dimension, so that D is a foliation of M. The usual procedure, as developed by Kostant, requires the assumption that the space of leaves M/D is a Hausdorff manifold. This may be too restrictive in practice, and it seems worthwhile to explore the possibility of using measures transverse to the foliation D, instead of densities on M/D, to construct the quantum phase space. To this end we outline the usual procedure for half-density quantisation as developed in (2) and relate it to Connes' construction of transverse measures.

WAVE FUNCTIONS

For each $m \in M$ we denote by T_m the tangent space to M at m and F_m the fibre of F at m. We write $D_m = F_m \cap \bar{F}_m \cap T_m$ and $E_m = (F_m + \bar{F}_m) \cap T_m$. Then $D_m^{\mathbb{C}} = F_m \cap \bar{F}_m$ and $E_m^{\mathbb{C}} = F_m + \bar{F}_m$ are the fibres of sub-bundles $D^{\mathbb{C}}$ and $E^{\mathbb{C}}$ respectively of $T^{\mathbb{C}}$. The sub-bundle $D^{\mathbb{C}}$ is involutive (i.e. closed under the Lie bracket of vector fields) and we assume that $E^{\mathbb{C}}$ is also.

For any real or complex vector space W and real number r we denote by $\Delta_r(W)$ the 1-dimensional complex vector space of complex valued r-densities on W. The symplectic bilinear form ω_m on T_m induces a symplectic bilinear form on $V_m = E_m/D_m$ whose 1-density we denote by $|V_m|$.

We have a direct sum

$$V_m^{\mathbb{C}} = F_m/D_m^{\mathbb{C}} \oplus \bar{F}_m/D_m^{\mathbb{C}}$$

so that $|V_m| = \lambda_m . \bar{\lambda}_m$ for a unique positive real 1-density

λ_m on $F_m/D_m^{\mathbb{C}}$. The natural isomorphism

$$\Delta_r(D_m) \otimes \Delta_r(F_m/D_m^{\mathbb{C}}) = \Delta_r(F_m)$$

defines a natural isomorphism

$$\Delta_r(D_m) = \Delta_r(F_m)$$

by $\mu_m \to \mu_m \otimes \lambda_m^r$.

Let E^o be the bundle whose fibre at m is the annihilator E_m^o of E_m in $(T^{\mathbb{C}})^*$. Sections of E^o are stable under the Lie derivative along sections of E, and the symplectic form gives a natural isomorphism of D with E^o. This defines a partial connection on D along sections of E, and hence a partial connection on $\Delta_r(D) = \Delta_r(F)$ along sections of E, for each r.

Now let M be quantisable and let L be a hermitian line bundle with connection having $\frac{2\pi}{h}\omega$ as curvature, where h is Planck's constant. Then we have a partial connection on $L \otimes \Delta_{-\frac{1}{2}}(F)$ along sections of E. A section of $L \otimes \Delta_{-\frac{1}{2}}(F)$ is called an F-wave function if it is convariant constant along the sections of F.

If ψ and ψ' are F-wave functions then in a neighbourhood of $m \in M$ we have

$$\psi = s \otimes \nu , \quad \psi' = s' \otimes \nu'$$

where s,s' are sections of L and ν,ν' are sections of $\Delta_{-\frac{1}{2}}(F)$. The direct sum $V_m^{\mathbb{C}} = F_m/D^{\mathbb{C}} \oplus \bar{F}_m/D_m^{\mathbb{C}}$ gives a natural isomorphism

$$\Delta_{-\frac{1}{2}}(F_m) \otimes \Delta_{-\frac{1}{2}}(\bar{F}_m) \otimes \Delta_{\frac{1}{2}}(V_n) = \Delta_{-1}(D_m)$$

Therefore we have a well-defined (-1)-density (ψ,ψ') on D given by

$$(\psi,\psi')_m = (s,s')_m \ \nu_m \ \bar{\nu}'_m \ |V_m|^{\frac{1}{2}}$$

TRANSVERSE MEASURE

Let G be the holonomy groupoid of the foliation D, as defined in (1) VII. Then each element of G is a morphism $\gamma: x \to y$ with $x, y \in M$. We write $s(\gamma) = x$ and $r(\gamma) = y$ and $G^y = \{\gamma \in G \mid r(\gamma) = y\}$. Then s is a covering map of G^y onto the leaf of D through y. Let α be a 1-density on D. We denote by $s^*\alpha$ the function which to each $y \in M$ associates the measure ν^y on G^y corresponding to the 1-density on G^y given by the 1-density α on D and the covering map $s: G^y \to D$. Then $s^*\alpha$ is a transverse function on G as defined in (1) I. The (-1)-density (ψ, ψ') on D defines a unique transverse measure $\langle\psi, \psi'\rangle$ on G such that

$$\langle\psi, \psi'\rangle (s^*\alpha) = \int_M (\psi, \psi')\, \alpha \, dm$$

where dm is the Liouville 1-density on M.

The integral of the transverse measure $\langle\psi, \psi'\rangle$ is proposed as the appropriate definition of the inner product of the F-wave functions ψ and ψ' in the case when the space of leaves M/D is not a Hausdorff manifold.

REFERENCES

(1) A. Connes. Sur la théorie non commutative de l'integration. Lecture Notes in Mathematics 725, pp. 19-143, Springer-Verlag, 1979.

(2) N. Woodhouse. Geometric quantization. Oxford University Press, 1980.

ON QUANTIZATION OF SYSTEMS WITH CONSTRAINTS

Jędrzej Śniatycki
Department of Mathematics and Statistics
The University of Calgary

Quantization of systems with constraints given by the vanishing of the momentum map of a gauge group is investigated. Applications to Yang-Mills fields are discussed.

Let (P,ω) be a (weakly) symplectic manifold, G a Lie group (possibly infinite dimensional), $\mathfrak{g}$ the Lie algebra of G, $\mathfrak{g}^*$ the dual of $\mathfrak{g}$, and $J\colon P \to \mathfrak{g}^*$ an equivariant momentum mapping corresponding to a Hamiltonian action of G in (P,ω). For each $\gamma \in \mathfrak{g}$, we denote by $J_\gamma\colon P \to \mathbb{R}$ the momentum corresponding to γ defined by

$$J_\gamma(p) = \langle J(p),\gamma\rangle \tag{1}$$

for every $p \in P$, where $\langle,\rangle$ denotes the evaluation. We consider (P,ω) to be the extended phase space of a dynamical system with constraints given by

$$J = 0. \tag{2}$$

A situation of this type appears in the temporal gauge formulation of the dynamics of Yang-Mills fields [1]. In the case of the structure group $SU(2)$, P is the space of the Cauchy data $(E_i^{\ a}, A_i^{\ a})$ for the field equations, which are of Sobolev class s and $s+1$, respectively, where $s \geq 1$, and G is the group of local gauge transformation

$$A_i^{\ a} \to g^{ab} A_i^{\ b} + \frac{1}{2}\,\varepsilon^{abc} g^{bd} \partial_i g^{cd} \tag{3}$$

$$E_i^{\ a} \to E_i^{\ b} g^{ba} \tag{4}$$

where $g = (g^{ab}(x))$ is a rotation matrix depending smoothly on x,

S. Sternberg (ed.), Differential Geometric Methods in Mathematical Physics, 207–211.

and equal to the identity matrix outside a compact set. For each $\gamma = (\gamma^{ab}(x))$ in $\mathcal{g}$, the corresponding momentum is given by

$$J_\gamma = \int (E_i^a A_i^b \gamma^{ab} - \frac{1}{2} \varepsilon^{abc} E_i^a \partial_i \gamma^{bc}) d_3 x. \tag{5}$$

There are two approaches to the canonical quantization of systems with constraints. One approach, suggested by P.A.M. Dirac [2] consists of a quantization of the extended phase space and a subsequent imposition of the quantum constraint condition. In the case when the constraints are given by the vanishing of the momentum map one can implement Dirac's approach as follows. Find a quantization of the extended phase space (P,ω) which leads to a representation of G, and required that the physically admissible states should be G invariant. It should be noted in the case when G is not compact the G invariant states will be given by generalized (non-normalized) vectors in the representation space. Thus, in this case one has to define a new scalar product in the space of the G invariant generalized vectors, which will describe the probability amplitudes for transitions between physically admissible states.

In the case of the Yang-Mills field the quantum constraint condition implies that the physically admissible states must be invariant under the group G of the local gauge transformations. The theory is also invariant under the group G_1 of the gauge transformations of the first kind which are given by equations (3,4) with constant matrices $g = (g^{ab})$. It can be shown in examples that the invariance of the physically admissible states under the group G_1 of the gauge transformations of the first kind depends on the topology of the representation space [3]. If this result extends to Yang-Mills fields interacting with matter fields, the freedom of the choice of the representation space might be of physical importance. With a choice of the representation space which allows for physical states which are not G_1 invariant one could have physical particles corresponding to the Yang-Mills fields. A choice of the representation space which leads to the G_1 invariance of the physical states could replace a confinement mechanism [4].

An alternative approach to the problem of a canonical quantization of a system with constraints consists of a quantization of the reduced phase space. There are several notions of reduction. The Marsden-Weinstein reduction of the constraint

$$J = 0 \tag{6}$$

applies to the case when 0 is a regular value of the momentum map J, and it yields the space $J^{-1}(0)/G$ of the G orbits in $J^{-1}(0)$ as the reduced phase space [5]. In this case one can show that, under certain regularity conditions, the quantization of the extended phase space and a subsequent imposition of the quantum constraint conditions yields results equivalent to the results obtained by the corresponding quantization of the reduced phase space [6].

In some cases of physical interest zero is not a regular value of the momentum map. For example, the momentum map for the $SU(2)$ Yang-Mills field given by equation (5) has quadratic singularities. It has been shown by J. Arms, J. Marsden and V. Moncrief that under some technical assumptions the singularities of the momentum map are at most quadratic [7].

In the case when zero is not a regular value of the momentum map it is not clear what is the appropriate reduction procedure. The following algebraic definition of reduction was proposed by A. Weinstein. Let J denote the ideal in the associative algebra structure of $C^{\infty}(P)$ which is generated by the components of the momentum map J, and $\rho\colon C^{\infty}(P) \to C^{\infty}(P)/J$ the canonical projection. The action of G in P induces actions of G in $C^{\infty}(P)$ and $C^{\infty}(P)/J$ such that ρ is G equivariant. Let A be the space of G invariant elements in $C^{\infty}(P)/J$. The Poisson algebra structure in $C^{\infty}(P)$ induces a Poisson algebra structure in A.

The Poisson algebra A is called the reduced Poisson algebra of the system. If zero is a regular value of J, and $J^{-1}(0)/G$ is a quotient manifold of $J^{-1}(0)$, then the reduced Poisson algebra is canonically isomorphic to the Poisson algebra of the reduced phase space. If 0 is not a regular value of J, then the reduced Poisson algebra need not correspond to a symplectic manifold. In this case the quantization by reduction should involve a quantization of the reduced Poisson algebra.

We have analysed a simple example where $P = \mathbb{R}^4$, $\omega = dp_1 \wedge dq_1 + dp_2 \wedge dq_2$, and G is the one parameter subgroup of $Sp(2,\mathbb{R})$ generated by $J = q_1{}^2 + q_2{}^2$. In this case the reduced Poisson algebra consists of formal power series

$$a(z,w) = \sum_{n=0} \frac{z^n}{n!} a_n(w) \tag{7}$$

where $a_n(w)$ are entire analytic functions of a complex variable w, and a_0 is a real constant. The Poisson bracket is given by

$$[a(z,w),b(z,w)] = \frac{\partial a}{\partial w}\frac{\partial b}{\partial z} - \frac{\partial a}{\partial z}\frac{\partial b}{\partial w} . \tag{8}$$

The reduced Poisson algebra can be quantized yielding results equivalent to the results of a quantization of the extended phase space and a subsequent imposition of the quantum constraint condition [8].

It would be very interesting to know the structure of the reduced Poisson algebra of the Yang-Mills theory. In this case the constraint $J^{-1}(0)$ has the regular component R consisting of the Cauchy data for which the solutions have no infinitesimal symmetries, and a singular component S corresponding to the solutions with symmetries,

$$J^{-1}(0) = R \cap S. \tag{9}$$

The regular component R is open and dense in $J^{-1}(0)$, and the space R/G of the G orbits in R has a weakly symplectic structure. The reduced Poisson algebra $\mathcal{A}$ contains the algebra of smooth functions on R/G. The question arises whether $\mathcal{A}$ contains any other elements; such elements would correspond to the G orbits contained in the singular component S of $J^{-1}(0)$. If $\mathcal{A}$ is equal to $C^{\infty}(R/G)$, then it is possible that one could ignore the singular component and obtain all the information about the quantum behaviour of the Yang-Mills field by quantizing R/G. On the other hand, if $\mathcal{A}$ contains elements corresponding to the G orbits in S in such a way that it is not isomorphic to the Poisson algebra of a symplectic manifold, then the quantization of the Yang-Mills field via reduction might require the quantization of the Poisson algebra $\mathcal{A}$ rather than the quantization of a symplectic manifold.

References

[1] Segal, I., J. Func. Anal., 33 (1979), p. 175; Eardlay, D.M. and Moncrief, V., Comm. Math. Phys., 83 (1982), pp. 193, 213.

[2] Dirac, P.A.M., Can. J. Math., 2 (1950), p. 129.

[3] Śniatycki, J., "On gauge invariance of physical states," Preprint No. 528, Department of Mathematics and Statistics, University of Calgary.

[4] Ito, H., Ann. Phys., 141 (1982), p. 290.

[5] Marsden, J.E. and Weinstein, A., Rep. Math. Phys., 5 (1974), p. 121.

[6] Tulczyjew, W.M., Bull. Acad. Polon. Sci., Sér. sci. math., astr. et phys., 13 (1965), p. 329; Guillemin, V. and Sternberg, S., "Geometric quantization and multiplicities of group representations" (to appear); Śniatycki, J., "Constraints and quantization," Preprint No. 506, Department of Mathematics and Statistics, University of Calgary.

[7] Arms, J., Marsden, J.E. and Moncrief, V., Comm. Math. Phys., 78 (1981), p. 455.

[8] Śniatycki, J. and Weinstein, A., "Reduction and quantization for singular momentum mappings," Lett. Math. Phys. (to appear).

GEOMETRIC QUANTIZATION IN THE SPIRIT OF GUPTA AND BLEULER

Joseph A. Wolf

Department of Mathematics
University of California
Berkeley, CA 94720

The standard method of geometric quantization produces the tempered representations of a semisimple Lie group. Here an extension of that method is described, which produces singular unitary representations. The method has strong formal similarities with Gupta-Bleuler quantization of the transverse photon.

Just to place the context of my talk, let me remind you that a semisimple Lie group has several kinds of unitary representations. The ones that enter into the Plancherel formula for the group are the ones whose characters are tempered distributions. These are very well understood now. For each conjugacy class of Cartan subgroups H in the semisimple Lie group G, there is a series of representations with a discrete parameter associated to the compact part of H and a continuous parameter associated to the noncompact part of H. See (5) and (10).

The delicate part of this is the case where H is compact (modulo the center of G). The analysis of those representations is Harish-Chandra's famous theory of the discrete series, which I denote $\hat{G}_{disc}$. See (3), (4) and (10). In fact, both for convenience and for technical reasons, one deals with a slightly larger class of groups, the reductive Lie groups. The discrete series picture for reductive groups is summarized in Plate I. Part 1 is the heart of the matter there, Parts 2 and 3 are just some necessary technicalities.

The series for the other conjugacy classes of Cartan sub-

*Research partially supported by N.S.F. Grant MCS-7902522.

S. Sternberg (ed.), Differential Geometric Methods in Mathematical Physics, 213–224.

PLATE I. Discrete Series

G: reductive Lie group, identity component G^0, Lie algebra $\mathfrak{g}_0$ such that (i) $Z_G(G^0)/Z_{G^0}$ is compact, (ii) $G/G^\dagger$ is finite, where $G^\dagger = Z_G(G^0)\cdot G^0$, and (iii) if $x \in G$ then Ad(x) is an inner automorphism on the complexified Lie algebra $\mathfrak{g} = \mathfrak{g}_0 \otimes_R C$.

T: compactly embedded Cartan subgroup of G, i.e. $\mathfrak{t}_0$ is a Cartan subalgebra of $\mathfrak{g}_0$, $T = Z_G(\mathfrak{t}_0)$, and $T/Z_G(G^0)$ is compact.

K: maximal compactly embedded subgroup of G that contains T, i.e. $Z_G(G^0) \subset K$ and $K/Z_G(G^0)$ is a maximal compact subgroup of the linear semisimple Lie group $G/Z_G(G^0)$.

W: Weyl group $W(G^0,T^0) = N_{G^0}(T)/T$.

Φ: root system of $\mathfrak{g}$ relative to $\mathfrak{t}$; Φ^+: positive roots.

ρ: half the sum of the positive roots.

G': regular elements of G, i.e. elements $x \in G$ such that the fixed point set of Ad(x) is a Cartan subalgebra of $\mathfrak{g}$.

Part 1: Topological Identity Component G^0.

$\widehat{G^0}_{disc} = \{\pi^0 \in \widehat{G^0} : \pi^0$ has coefficients in $L_2(G^0/Z_{G^0})\}$ has an element π^0_λ, for every nonsingular integral $\lambda \in i\mathfrak{t}^*$, whose distribution character is given on $T \cap (G^0)'$ by the formula $\Theta(\pi^0_\lambda) = \pm\left\{\sum_{w\in W} \mathrm{sign}(w)e^{w\lambda}\right\}/\prod(e^{\alpha/2} - e^{-\alpha/2})$. The π^0_λ exhaust $\widehat{G^0}_{disc}$, $\pi^0_\lambda = \pi^0_{\lambda'}$ if and only if $\lambda \in W(\lambda')$, and π^0_λ has infinitesimal character χ_λ.

Part 2: Algebraic Identity Component $G^\dagger = Z_G(G^0)\cdot G^0$.

$\widehat{G^\dagger}_{disc} = \{\pi^\dagger_{\lambda,\psi} : \psi \in Z_G(G^0)^\wedge$ agrees with $e^{\lambda-\rho}$ on $Z_{G^0}\}$ where $\pi^\dagger_{\lambda,\psi}(zg) = \psi(z) \otimes \pi^0_\lambda(g)$ for $z \in Z_G(G^0)$ and $g \in G^0$. $\pi^\dagger_{\lambda,\psi}$ has infinitesimal character χ_λ and has distribution character $\Theta(\pi^\dagger_{\lambda,\psi})(zg) = \{\mathrm{trace}\ \psi(z)\} \times \Theta(\pi^0_\lambda)(g)$.

Part 3: The Entire Group G.

$\hat{G}_{disc} = \{\pi_{\lambda,\psi} = \mathrm{Ind}^G_{G^\dagger}(\pi^\dagger_{\lambda,\psi})\}$. $\pi_{\lambda,\psi}$ has infinitesimal character χ_λ and has distribution character $\Theta(\pi_{\lambda,\psi}) = 0$ on $G\backslash G^\dagger$,

$$\Theta(\pi_{\lambda,\psi})(zg) = \sum \Theta(\pi^\dagger_{\lambda,\psi})(x_i^{-1} zgx_i) \text{ on } G^\dagger, \text{ where } G = \cup x_i G^\dagger .$$

groups are constructed from the discrete series of certain special subgroups. Any Cartan subgroup has a natural splitting $H = T \times A$ where $T \subset K$ is the compact part and where $A = \exp(\mathfrak{a}_0)$ is a vector group that is orthogonal to K in a suitable sense: roots are pure imaginary on $\mathfrak{t}_0$ and real on $\mathfrak{a}_0$. Restrict the roots to $\mathfrak{a}_0$ to get the $\mathfrak{a}_0$-roots of $\mathfrak{g}_0$, pick a positive subsystem, let $\mathfrak{n}_0$ be the sum of the positive $\mathfrak{a}_0$-root spaces, and you have a cuspidal parabolic subgroup

$$P = MAN \quad , \qquad MA = M \times A = Z_G(A) \quad .$$

M satisfies the reductive group conditions for G on Plate I, and T is a compactly embedded Cartan subgroup of M. The corresponding series of unitary representations,

$$\{\mathrm{Ind}_P^G(\eta \otimes e^{i\mu}): \quad \eta \in \hat{M}_{disc} \text{ and } \mu \in \mathfrak{a}_0^*\} \quad ,$$

depends only on the conjugacy class of $H = T \times A$. My only point here is that the parameters of η provide a discrete parameter for this series, and μ is a continuous parameter.

What about the other unitary representations? There certainly are many non-tempered representations in general. One gets some by letting the continuous parameter go non-real inside $\mathfrak{a} = \mathfrak{a}_0 \otimes_R C$. I'll discuss the ones obtained by letting η go singular, in other words by dropping the nonsingularity condition on λ in the parametrization (Plate I) of the discrete series of M.

Now let us concentrate on the key case: continuation of the discrete series when G is connected.

Specifically, I want to describe the early stages of a uniform geometric construction of unitary representations, which includes the geometric construction of the discrete series. This represents completed joint work with John Rawnsley and Wilfried Schmid and continuing joint work with Wilfried Schmid. See (8). In the language of geometric quantization, it may lead to quantization of all elliptic co-adjoint orbits, in particular to a geometric treatment of all positive energy (= lowest weight) representations. The basic setup is given in Plate II. That picture, and the point of looking in dimension s, is motivated by the classical case, which is recalled in Plate III. In that classical case, note that (2) really is the Kostant-Langlands Conjecture, and that given a discrete series representation one can choose a positive root system Φ^+ so that the representation does occur on harmonic forms of degree s.

As the representation goes singular, its coefficients grow faster, and we cannot hope to find it on a space of square integrable forms on G/(compact) for any finite dimensional vector bundle. So it is natural to try to imitate the classical procedure

PLATE II. Setup for the Elliptic Case

G: connected relative Lie group, e.g. $U(k,\ell)$.

H: centralizer of a torus subgroup, e.g. $U(k_1,\ell_1) \times U(k_2,\ell_2)$ inside $U(k_1+k_2, \ell_1+\ell_2)$.

$\mathfrak{g}_0, \mathfrak{h}_0$: respective real Lie algebras.
$\mathfrak{g}, \mathfrak{h}$: complexified Lie algebras.

Invariant complex structures on G/H are in one-one correspondence with parabolic subalgebras $\mathfrak{h} + \mathfrak{n}_-$ of $\mathfrak{g}$ with reductive part $\mathfrak{h}$; here $\mathfrak{n}_+ = \overline{\mathfrak{n}_-}$ represents the holomorphic tangent space of G/H. Example: $G/H = U(k_1+k_2, \ell_1+\ell_2)/U(k_1,\ell_1) \times U(k_2,\ell_2)$ where $G_{\mathbb{C}} = GL(k_1+k_2 + \ell_1+\ell_2; C)$ has Lie algebra given in matrix block form as indicated here. There are two invariant complex structures; interchange $\mathfrak{n}_+$ and $\mathfrak{n}_-$ to obtain the one from the other. They are realized as open G-orbits on the complex flag manifold $G_{\mathbb{C}}/H_{\mathbb{C}}Q_-$.

$$\begin{pmatrix} \mathfrak{h} & \mathfrak{n}_+ & \mathfrak{h} & \mathfrak{n}_+ \\ \mathfrak{n}_- & \mathfrak{h} & \mathfrak{n}_+ & \mathfrak{h} \\ \mathfrak{h} & \mathfrak{n}_- & \mathfrak{h} & \mathfrak{n}_+ \\ \mathfrak{n}_- & \mathfrak{h} & \mathfrak{n}_- & \mathfrak{h} \end{pmatrix} \begin{matrix} \}\, k_1 \\ \}\, k_2 \\ \}\, \ell_1 \\ \}\, \ell_2 \end{matrix}$$
$$\quad \underbrace{}_{k_1} \quad \underbrace{}_{k_2} \quad \underbrace{}_{\ell_1} \quad \underbrace{}_{\ell_2}$$

ψ: irreducible unitary representation of H, i.e. $\psi \in \hat{H}$.
V: representation space of ψ.
$\mathbf{V} \to G/H$: associated G-homogeneous holomorphic vector bundle with invariant hermitian metric derived from V.
s: complex dimension of the maximal compact subvariety K/L, $L = K \cap H$, of G/H.

$H^p(G/H,\mathbf{V})$: Dolbeault cohomology in degree p.

$\tilde{\mathcal{H}}_2^p(G/H,\mathbf{V})$: "cohomology" from L harmonic forms.

The Problem:

Define the space $\tilde{\mathcal{H}}_2^s(G/H,\mathbf{V})$ so that it is a Hilbert space, understand the unitary representation of G on $\tilde{\mathcal{H}}_2^s(G/H,\mathbf{V})$, and use that to unitarize the Fréchet representation of G on $H^s(G/H,\mathbf{V})$.

PLATE III. Classical Case.
Schmid's Solution to the Kostant-Langlands Conjecture.

G and H are as before, but with H compact. Fix a (positive definite) G-invariant hermitian metric on G/H. Choose a compact Cartan subgroup T of G which is contained in K. As before, Φ denotes the root system of $(\mathfrak{g}_0, \mathfrak{t}_0)$. Let Φ^+ denote a positive root system such that $\mathfrak{n}_+ = \sum_{\alpha \in \Phi^+ \setminus \Phi^+(\mathfrak{h})} \mathfrak{g}_\alpha$ represents the holomorphic tangent space. ψ, V and $\mathbf{V}$ are as before, with ψ irreducible, thus finite dimensional.

Define:

χ: highest weight of the irreducible representation ψ.

$L_2^p(G/H,\mathbf{V}) = \{\mathbf{V}\text{-valued } (0,p)\text{-forms } \phi\colon \int_{G/H} \|\phi(x)\|^2 dx < \infty\}$

$\bar\partial^*$: formal adjoint of $\bar\partial\colon L_2^p(G/H,\mathbf{V}) \longrightarrow L_2^{p+1}(G/H,\mathbf{V})$.

$\square = \bar\partial\bar\partial^* + \bar\partial^*\bar\partial$, complex Laplace-Beltrami operator.

$\mathcal{H}_2^p(G/H,\mathbf{V}) = \{\phi \in L_2^p(G/H,\mathbf{V})\colon \square\phi = 0 \text{ as distribution}\}$.

Theorems:

1. If $\chi+\rho$ is singular then every $\mathcal{H}_2^p(G/H,\mathbf{V}) = 0$.
2. Suppose that $\chi+\rho$ is nonsingular and define

   $q(\chi+\rho) = \#\{\text{compact positive roots } \alpha\colon (\alpha, \chi+\rho) < 0\}$
   $\qquad + \#\{\text{noncompact positive roots } \beta\colon (\beta, \chi+\rho) > 0\}$.

   If $p \neq q(\chi+\rho)$ then $\mathcal{H}_2^p(G/H,\mathbf{V}) = 0$;
   if $p = q(\chi+\rho)$ then $\mathcal{H}_2^p(G/H,\mathbf{V}) \neq 0$, and G acts irreducibly on it by the discrete series representation $\pi_{\chi+\rho}$.
3. Suppose that $\chi+\rho$ is nonsingular and $s = q(\chi+\rho)$.
   Then the map $\mathcal{H}_2^s(G/H,\mathbf{V}) \to H^s(G/H,\mathbf{V})$, of a harmonic form to its Dolbeault class, is an isomorphism on the K-finite level; so $\mathcal{H}_2^s(G/H,\mathbf{V})$ unitarizes $H^s(G/H,\mathbf{V})$.

on a bundle $\mathbf{V} \to G/H$ where H need not be compact and $\mathbf{V}$ need not be finite dimensional.

What are the difficulties in imitating the classical procedure, when H is noncompact? In order to answer that I must be more specific about just what we are trying to do.

PLATE IV. Specific Program

θ: Cartan involution of G such that $\theta(H) = H$.
K: maximal compactly embedded subgroup of G given by $K = G^{\theta}$.
$L = H \cap K$, so K/L is a maximal compact subvariety of G/H and s is the complex dimension of K/L.
$\langle\ ,\ \rangle$: G-invariant indefinite-Kaehler metric on G/H.

Problems:

1. Define an auxiliary K-invariant G-bounded positive definite hermitian metric on G/H, and define the space $\mathcal{H}_2^s(G/H,\mathbf{V})$ of $\mathbf{V}$-valued $(0,s)$-forms on G/H that are L_2 for the positive definite metric and harmonic for the invariant metric.

2. Show that the G-invariant global hermitian form $\langle \phi,\phi' \rangle_{G/H} = \int_{G/H} \langle \phi(x),\phi'(x)\rangle\, dx$ is semidefinite on $\mathcal{H}_2^s(G/H,\mathbf{V})$.

3. Show that the natural map $\mathcal{H}_2^s(G/H,\mathbf{V}) \to H^s(G/H,\mathbf{V})$ of a harmonic form to its Dolbeault class is surjective on the K-finite level.

4. Show that the kernel of $\langle\ ,\ \rangle_{G/H}$ on $\mathcal{H}_2^s(G/H,\mathbf{V})$ coincides with the kernel of the natural map to Dolbeault cohomology.

If all this goes through:
then the action of G on $\mathcal{H}_2^s(G/H,\mathbf{V})$ induces a unitary representation of G on $\mathcal{H}^s(G/H,\mathbf{V})/(\text{kernel of } \langle\ ,\ \rangle_{G/H})$ which unitarizes the Fréchet representation of G on $H^s(G/H,\mathbf{V})$.

Note the similarity to the Gupta-Bleuler quantization scheme. This was pointed out to me separately by Flato, Fronsdal and Varadarajan after they heard about this work. The kernel corresponds to the longitudinal photons, and the quotient corresponds to the space of transverse photons. The scalar photons also have an analog here — we'll see it later.

Several conditions are necessary before a program like this can have any hope of success.

1. The notion of L_2 should be canonical and well defined. First, the auxiliary positive definite hermitian metric must be K-invariant so that we can keep track of K-types. Second, even though a general element of G distorts L_2-norm, it should be bounded on any closed space of L_2 forms. Third, the global invariant hermitian form $\langle\ ,\ \rangle_{G/H}$ should be jointly continuous on any of those Hilbert spaces. Wilfried Schmid and I have carried this out in general.

2. The notion of "harmonic" must be clarified. There really are two choices,

$$(\bar\partial\phi = 0 \quad \text{and} \quad \bar\partial^*\phi = 0) \quad \text{and} \quad (\bar\partial\bar\partial^* + \bar\partial^*\bar\partial)\phi = 0 \quad .$$

We use the first because we need to be able to compare our harmonic spaces with Dolbeault cohomology spaces. That comparison comes into the unitarization procedure itself, and we also use it to identify the resulting representations and prove that they are irreducible. See item 3 just below.

3. We must understand the Fréchet representation, say π_V, of G on $H^s(G/H,\mathbf{V})$. Ideally this means that we should show that π_V is admissible, we should find its infinitesimal, distribution and K-characters, and we should work out a concrete description of its K-spectrum. Schmid and I have carried this out in a moderately general setting.

4. The indefinite-Kaehler geometry of G/H should be related to the Kaehler geometry of K/L so that we can understand what it means for a form on G/H to be harmonic. Schmid and I have done this in a somewhat restricted context.

The key to #3 and #4 is a fibration $\pi: G/H \to K/L$ and a variation on the Leray spectral sequence. See Plate V, next page. When the fibre V of $\mathbf{V} \to G/H$ is finite dimensional, this reduces analysis of the K-spectrum of $H^s(G/H,\mathbf{V})$ to an algebraic question, and when $\dim V = 1$ the algebraic question is easily answered and it shows that $H^s(G/H,\mathbf{V})$ is K-multiplicity free.

More generally, Schmid and I get character information by resolving the vector bundle $\mathbf{V} \to G/H$ and using methods of coherent continuation. The main results for the finite dimensional case are collected in Plate VI. These character formulae are completely explicit when $\mathbf{V} \to G/H$ is finite dimensional and hermitian and $\pi: G/H \to K/L$ is holomorphic; see (6).

PLATE V. The Fibration and the Spectral Sequence

The Fibration:

$\mathfrak{g}_0 = \mathfrak{h}_0 + \mathfrak{q}_0$ where $\mathfrak{q}_0 = (\mathfrak{q}_+ + \mathfrak{q}_-) \cap \mathfrak{g}_0$

$\mathfrak{g}_0 = \mathfrak{k}_0 + \mathfrak{p}_0$ where $\mathfrak{p}_0$ is the (-1)-eigenspace of θ on $\mathfrak{g}_0$.

Theorem of Mostow (7):

$(k,\xi,\eta) \to k\cdot\exp(\xi)\cdot\exp(\eta)$ defines a diffeomorphism of $K \times (\mathfrak{p}_0 \cap \mathfrak{q}_0)\times(\mathfrak{p}_0 \cap \mathfrak{z}_0)$ onto G.

Reformulation of Mostow's Theorem:

$\pi(k\cdot\exp(\xi)\cdot\exp(\eta)) = kL \quad (k\in K,\ \xi \in \mathfrak{p}_0 \cap \mathfrak{q}_0,\ \eta \in \mathfrak{p}_0 \cap \mathfrak{h}_0)$ defines a C^∞ fibre bundle $\pi\colon G/H \to K/L$ with fibre $F = \mathfrak{p}_0 \cap \mathfrak{q}_0$ and with structure group L acting on F by restriction of the adjoint representation of G.

The Spectral Sequence:

1. Suppose $\mathfrak{p}_0 \cap \mathfrak{q}_0 = \dot{\mathfrak{m}}_0 \cap \mathfrak{p}_0$ for some θ-stable subalgebra $\mathfrak{m}_0$ of $\mathfrak{g}_0$. Then F is a bounded symmetric domain holomorphically embedded in G/H, say F = M/L.

2. Further, there is a spectral sequence abutting to $H^*(G/H,\mathbf{V})$ with $E_2^{p,q} = E_\infty^{p,q} = H^p(K/L, \mathbf{H}^q(M/L,\mathbf{V}))$ on the K-finite level. If G/H is symmetric it gives

(*) $$H^s(G/H,\mathbf{V})_K = H^s_\delta(K/L, \mathbf{H}^0(M/L,\mathbf{V})_L)_K \quad \text{as a K-module,}$$

where δ is a certain first order differential operator on the bundle $\mathbf{H}^0(M/L,\mathbf{V}) \to K/L$.

3. If G/H is symmetric and $\pi\colon G/H \to K/L$ is holomorphic, then $\mathbf{H}^0(M/L,\mathbf{V}) \to K/L$ is holomorphic, δ reduces to its $\bar\partial$-operator, and (*) gives the K-spectrum of $H^s(G/H,\mathbf{V})$ by means of the Bott-Borel-Weil Theorem.

PLATE VI. Character Formulae: Case Rank G = rank K

If ψ is finite dimensional but not necessarily unitary:

Let $\mathcal{C}$: negative Weyl chamber in $i\mathfrak{t}_0^*$

$\Theta(\mathcal{C},\lambda)$: the coherent family of invariant eigendistributions on G such that, if λ is regular, then $\Theta(\mathcal{C},\lambda)$ is the character of the discrete series representation π_λ.

χ: heighest weight of ψ.

Then

$$\sum_{p\geqq 0}(-1)^p\,\Theta(H^p(G/H,\mathbf{V})) = \frac{(-1)^{\dim K/T}}{|W_L|}\sum_{u\in W(H)}\det(u)\Theta(\mathcal{C},\,u(\chi+\rho))$$

and

$$|W_L|\sum_{p\geqq 0}(-1)^p\Theta_K(H^p(G/H,\mathbf{V})) = \sum_{\substack{n_i\geqq 0\\ u\in W(H)}}\frac{\sum_{w\in W(K)}\det(w)\,e^{w\left(u(\chi+\rho)-\rho+\rho_K-\Sigma n_i\beta_i\right)}}{\prod_{\alpha>0\text{ compact}}(e^{\alpha/2}-e^{-\alpha/2})}$$

If $\dim\psi<\infty$, ψ unitary, $\pi\colon G/H\to K/L$ holomorphic:

Each $H^p(G/H,\mathbf{V})$ is a Harish-Chandra module of finite type, T-finite with weights bounded from above, infinitesimal character $\chi_{\chi+\rho}$. Write Σ_v for the sum over the set of all v in $W(\mathfrak{h})$ such that $v(\rho)-\rho$ is L-dominant. Then the $\Theta(\mathcal{C},\lambda)$ are just holomorphic characters and the above formulae reduce to

$$\sum_{p\geqq 0}(-1)^p\,\Theta(H^p(G/H,\mathbf{V})) = \frac{(-1)^{\dim K/T}}{|W_L|}\sum_v\det(v)\;\Theta(v(\chi+\rho))$$ and

$$\sum_{p\geqq 0}(-1)^p\Theta_K(H^p(G/H,\mathbf{V})) = \sum_v\det(v)\Theta_K(v(\chi+\rho)) =$$

$$\sum_v\det(v)\sum_{n_i\geqq 0}\frac{\sum_{w\in W(K)}\det(w)\,e^{w\left(v(\chi+\rho)-\rho+\rho_K-\Sigma n_i\beta_i\right)}}{\prod_{\alpha>0\text{ compact}}(e^{\alpha/2}-e^{-\alpha/2})}$$

We now make the working assumption that G/H is symmetric (so, in particular, rank K = rank G here), that $V \to G/H$ is hermitian (i.e. ψ is unitary), and that $\pi: G/H \to K/L$ is holomorphic. Then it is easy, at least when V is finite dimensional, to follow equation (*) on Plate V, K-type by K-type, and the character theory described in Plate VI is explicit. In any case, one can follow square integrability through the spectral sequence, and the notion of "harmonic" on G/H becomes transparent. Thus, under certain negativity conditions on V which I will describe in a moment, we carry out the program described in Plate IV and produce irreducible, possibly singular, unitary representations in a uniform geometric manner.

Write $A^p(G/H,\mathbf{V})$ for the space of $\mathbf{V}$-valued C^∞ (0,p)-forms on G/H. We call a form $\phi \in A^p(G/H,\mathbf{V})$ horizontal if it is horizontal relative to $G/H \to K/L$, i.e. if $\phi(KM) \subset V \otimes \Lambda^p(\mathfrak{k} \cap \mathfrak{n}_-)^*$.

Theorem. A form $\omega \in A^s(K/L,\mathbf{H}^0(M/L,\mathbf{V}))_K$ is a harmonic on K/L if and only if the corresponding ((*) in Plate V) horizontal form $\phi \in A^s(G/H,\mathbf{V})$ is harmonic relative to the invariant indefinite Kaehler metric on G/H.

Since G/H is symmetric, H is the fixed point set of an involutive automorphism τ of G, and τ commutes with the Cartan involution θ defining K because $\theta(H) = H$. So the group M and its Lie algebra $\mathfrak{m}_0$, in #1 in the description (Plate V) of the fibration and spectral sequence, are the respective fixed point sets of $\theta\tau$ on G and $\mathfrak{g}_0$. Let $\{\gamma_i\}$ be the maximal roots of the noncompact simple factors of $\mathfrak{m}_0$. Consider the

L_2 ***Condition***: if $\nu \in \hat{L}$ and $V_\nu \neq 0$ then $(\nu + \rho_M, \gamma_i) < 0$ for all i.

Theorem. If the L_2 condition holds, then every class $c \in H^s(G/H,\mathbf{V})_K$ has a unique horizontal L_2 harmonic representative.

Notes. Modulo tensor factors of V corresponding to subgroups of H that act trivially on G/H, the L_2 condition forces V to have a highest L-type. If V has a highest L-type, say χ, then the L_2 condition reduces to: $(\chi + \rho_M, \gamma_i) < 0$ for all i.

Theorem. Suppose that, whenever $\nu \in \hat{L}$ and $V_\nu \neq 0$, we have

(**) $$2(\nu + \rho_M, \gamma_i)/(\gamma_i,\gamma_i) \leqq -1 \quad \text{for all } i .$$

Then $(-1)^s\langle\ ,\ \rangle_{G/H}$ is positive semidefinite on $\mathcal{H}_2^s(G/H,\mathbf{V})$, its null space on $\mathcal{H}_2^s(G/H,\mathbf{V})$ coincides with the kernel of the natural map $\mathcal{H}_2^s(G/H,\mathbf{V}) \to H^s(G/H,\mathbf{V})$, and that natural map is surjective on the K-finite level. In consequence, G acts on $\mathcal{H}_2^s(G/H,\mathbf{V})/$ (kernel $\langle\ ,\ \rangle_{G/H}$) by a unitary representation, and this unitarizes the Fréchet representation of G on $H^s(G/H,\mathbf{V})$.

Notes. If G, or even just M, is a linear group, or if V is finite dimensional, then (**) reduces to the L_2 condition. In any case one can get an almost identical but slightly less geometric result, with $\leqq -1$ in (**) weakened to $\leqq -\frac{1}{2}$.

Theorem: If V is finite dimensional, and π_V is the unitary representation constructed just above, then π_V is irreducible and its characters are explicitly described in Plate VI.

Let me indicate how this works in a very special case, the ladder representations of the conformal group. So G is the double cover of U(2,2), H is the subgroup of G that covers $U(1) \times U(1,2)$, $K/L = \{U(2) \times U(2)\}/\{U(1) \times U(1) \times U(2)\} = U(2)/U(1) \times U(1)$ is the complex projective line (Riemann sphere), $s = 1$, and by using the theorems just above with various negative holomorphic line bundles $\mathbf{V} \to G/H$ we obtain all but one of the ladder representations. See (8), Section 13. The ladder representation not obtained this way is the one with a 1-dimensional K-type. It is the very singular representation of SO(2,4) that remains irreducible on SO(1,4), and is associated to a nilpotent coadjoint orbit rather than an elliptic orbit. See (9) and (1).

Finally, let us return to quantum electrodynamics, specifically to the photon. Fronsdal and others have studied the notion Gupta-Bleuler Triple in the context of QED. That is an indecomposable but reducible representation $0 = X_0 \subset X_1 \subset X_2 \subset X_3 = X$ where X is an indefinite-unitary representation space, X_1 is a totally isotropic invariant subspace, and X_2 is the orthogonal of X_1. See (2) and (11). The scalar product on X pairs the representations X/X_2 and X_1, which turn out to be unitary in the cases described by Fronsdal. Thus, in the photon case, X/X_2 gives the scalar photon, X_2/X_1 gives the transverse photon, and X_1 gives the longitudinal photon. In the material I describe here, the setting is really different but the spirit is similar. Thus the analog of X is the weak harmonic space

$$\{\phi \in L_2^s(G/H,\mathbf{V}): \Box\phi = 0 \text{ in the sense of distributions}\} ,$$

the analog of X_2 is the harmonic space $\mathcal{H}_2^s(G/H,\mathbf{V})$ studied here, and the analog of X_1 consists of the $\bar{\partial}$-exact forms in $\mathcal{H}_2^s(G/H,\mathbf{V})$. The quotient representations I have described then correspond to the quantization of the transverse photon, the other two pieces to the scalar and longitudinal photon. But so far there is no general argument in the holomorphic setting that the invariant indefinite global hermitian form is nondegenerate on the weak harmonic space.

REFERENCES

(1) Blattner,R.J. and Wolf,J.A., "Explicit quantization of the Kepler manifold," Proc. Amer. Math. Soc., Vol. 77 (1979) pp.145-149.

(2) Fronsdal,C. "Semisimple gauge theories and conformal gravity," Lectures in Applied Mathematics, Proceedings of the 1982 Summer Seminar, American Mathematical Society, to appear.

(3) Harish-Chandra, "Discrete series for semisimple Lie groups, I," Acta Math., Vol. 113 (1965), pp.241-317.

(4) Harish-Chandra, "Discrete series for semisimple Lie groups, II," Acta Math., Vol. 116 (1966), pp.1-111.

(5) Harish-Chandra, "Harmonic analysis on semisimple Lie groups," Bull. Amer. Math. Soc., Vol. 76 (1970), pp.529-551.

(6) Hecht,H., "The characters of some representations of Harish-Chandra," Math. Ann., Vol. 219 (1976), pp.213-226.

(7) Mostow,G.D., "Some new decomposition theorems for semisimple Lie groups," Mem. Amer. Math. Soc., Vol. 14 (1955), pp.31-54.

(8) Rawnsley,J., Schmid,W. and Wolf,J.A., "Singular unitary representations and indefinite harmonic theory," Jour. Funct. Anal., Vol. 48 (1983), to appear.

(9) Sternberg,S. and Wolf,J.A., "Hermitian Lie algebras and metaplectic representations," Trans. Amer. Math. Soc., Vol. 238 (1978), pp.1-43.

(10) Wolf,J.A., "The action of a real semisimple Lie group on a complex flag manifold, II: Unitary representations on partially holomorphic cohomology spaces," Mem. Amer. Math. Soc., Vol. 138 (1974).

(11) Wolf,J.A., "Indefinite harmonic theory and unitary representations," Lectures in Applied Mathematics, Proceedings of the 1982 Summer Seminar, American Mathematical Society, to appear.

ON DEFORMATION OF DIFFERENTIALS OF IMMERSIONS

E. Binz and Th. Peter

Department of Mathematics
University of Mannheim

Abstract: Given a scalar product $<,>$ on $\mathbb{R}^n$ and a fixed smooth immersion i of a compact m-dimensional C^∞-manifold M into $\mathbb{R}^n$, the differential dj of any immersion j in the connected component of i is described relative to i as the pointwise formed composition $dj = g \cdot di \cdot f$ with a bundle isomorphism f of the tangent bundle TM, representing the Riemannian metric $j^*<,>$, and a smooth map g of M into the group $SO(n)$, called an integrating factor for $di \cdot f$. With this formalism we describe the Levi-Cività connection of $j^*<,>$, associated connections, torsions, second fundamental tensor and use these notions to reformulate the Dirac operator. Finally we describe the motion of an incompressible fluid.

S. Sternberg (ed.), Differential Geometric Methods in Mathematical Physics, 225–240.

o) Introduction

The purpose of this note is to present a formalism which describes the deformation of the differential of an immersion i mapping a compact smooth manifold M into an Euclidean space $(\mathbb{R}^n, <,>)$.

This formalism was developed to some extent in [5] mainly in order to lift Einstein's evolution equation from the space of Riemannian metrics of a three dimensional smooth manifold M onto the space of the differentials of all smooth immersions from M into an Euclidean space of a high codimension. A detailed and extended version of this formalism can be found in [12] , some preliminary treatments in [3] and [4] .

The basic idea is, that the space of smooth immersions from M into $\mathbb{R}^n$ forms an open set in the locally convex space of all smooth $\mathbb{R}^n$-valued functions of M . Thus if two embeddings i and j are near enough, they can be joined by an arc, which in turn yields, via standard arguments in homotopy theory, an isomorphism of their tangent - and normal bundles. This isomorphism relates the differentials $di, dj : TM \longrightarrow \mathbb{R}^n$ by a formula (which also holds for immersions) of the form

$$dj = g \cdot di \cdot f \ .$$

The two factors $f : TM \longrightarrow TM$, a bundle map symmetric and positive definite with respect to the pull back $i^*<,>$ of $<,>$ by i, and the smooth map $g: M \longrightarrow SO(n)$ play the following role: f, more precisely f^2, determines the metric $j^*<,>$ via $i^*<,>$. The factor g turns the $\mathbb{R}^n$-valued one form $di \cdot f$ into a differential, is hence an integrating factor. Its effect to the geometry is e.g. that it determines the curvature tensor of $j^*<,>$ up to a gauge transformation by f.

After the introduction of g and f and some basic investigations on covariant derivatives, torsions and second fundamental tensors we illustrate the use of the formalism on two examples:

First we suppose that M carries a spin structure and describe the change of the Dirac operator and the Laplace-Beltrami operator while we pass from i to j. Here we follow a treatment of R.Pferschy in [13]. Second we describe the equation of motion of a film of perfect fluid, which forms a moving hypersurface in $\mathbb{R}^n$, diffeomorphic to M. The motion is governed by its kinetic energy, only.

As references in differential geometry we use [8] and [11] and refer to [9] and [10] for the notion of smoothness in infinite dimensions.

1) The Relative Description of Immersions

Throughout these notes <,> is a fixed scalar product on $\mathbb{R}^n$ and M is a smooth compact manifold of dimension m . An immersion j from M into $\mathbb{R}^n$ is a smooth map of maximal rank everywhere. j is called an embedding if it is injective.

Assume $m < n$. The collection of immersions from M to $\mathbb{R}^n$ is denoted by $I(M,\mathbb{R}^n)$; the subset which consists of all embeddings is called $E(M,\mathbb{R}^n)$. Both sets are open in the Fréchet space $C^\infty(M,\mathbb{R}^n)$ of all smooth functions from M to $\mathbb{R}^n$, carrying Whitney's C^∞-topology.

Fix now an immersion i . Within its connected component $O_i \subset I(M,\mathbb{R}^n)$ any two maps can be joined by a smooth arc. Hence i is isotopic to any other $j \in O_i$. The smooth tangential representation $\tilde{j}$ of j assignes to any $p \in M$ the tangent plane $dj(p)(T_pM)$, regarded as an element in the Grassmanian $G(m,n)$ of all m-planes in $\mathbb{R}^n$. By dj we mean the principal part of the tangent map $Tj : TM \longrightarrow \mathbb{R}^n \times \mathbb{R}^n$.

The description of dj by di is achieved as follows: Denote by ξ and η the canonical m-resp.(n-m)-plane bundle. Their direct sum is the trivial bundle $G(m,n) \times \mathbb{R}^n$. Since i and j are isotopic the pullbacks $\tilde{i}^*\xi$ and $\tilde{j}^*\xi$ as well as $\tilde{i}^*\eta$ and $\tilde{j}^*\eta$ are isomorphic. ($\tilde{j}^*\xi \simeq TM$ and $\tilde{j}^*\eta$ is isomorphic to the normal bundle $\nu(j)$, regarded as a bundle over M.) Thus there is a smooth bundle isomorphism

$$B : \tilde{i}^*(\xi \oplus \eta) \longrightarrow \tilde{j}^*(\xi \oplus \eta)$$

preserving the direct sum. Both domain and range are canonically isomorphic to $M \times \mathbb{R}^n$ and thus B is described by a smooth map $\psi : M \longrightarrow GL(n)$. Choose B such that ψ satisfies $dj(p)(v_p) = \psi(p)(di(p)(v_p))$ for all $v_p \in T_pM$ and all $p \in M$. In short we have

$$dj = \psi \cdot di .$$

Now let $g \cdot \varphi$ be the polar decomposition of ψ formed pointwise with respect to $<,>$ on $\mathbb{R}^n$, where $g \in C^\infty(M,SO(n))$ and $\varphi \in C^\infty(M,GL(n))$. The latter is pointwise symmetric and positive definite. Based on the next observation the description $dj = g \cdot \varphi \cdot di$ will be refined as follows: The metrics $m(i)$ and $m(j)$, the pullbacks of $<,>$ by i and j, are related by

$$m(i)(AX,Y) = m(j)(X,Y) = <\varphi^2 \cdot diX, diY>$$

for a unique smooth bundle map $A : TM \longrightarrow TM$ and any two smooth vector fields X,Y on M. The map A is (pointwise) positive definite with respect to $m(i)$ and hence admits a unique positive square root $f : TM \longrightarrow TM$, again formed pointwise. Hence we have for each $v_p \in T_pM$ and each $p \in M$

$$dj(p)(v_p) = g(p)(di(p)(f(p)(v_p))) ,$$

or in short

$$dj = g \cdot di \cdot f .$$

This is the relative description of dj by di, in which $m(j)$ is reflected by the symmetric factor f. The role of the orthogonal factor g will be apparent later. In contrast to the uniqueness of the symmetric factor f the orthogonal one g is not unique: For any $q \in C^{\infty}(M,SO(n))$ which is the identity on $\tilde{i}^{*}\xi$ we have obviously

$$dj = g \cdot q \cdot di \cdot f .$$

Call any two $g_1, g_2 \in C^{\infty}(M,SO(n))$ to be equivalent if $(g_1 \cdot g_2^{-1}) | \tilde{i}^{*}\xi = id_{\tilde{i}^{*}\xi}$. Then $\bar{g}$, the equivalence class of g, together with f determine dj uniquely. Clearly $\bar{g}$ can be regarded as a smooth section into a fibre bundle over M with $SO(n)/_{SO(n-m)}$ as typical fibre. In case of $n = m+1$ and M oriented g is uniquely determined by the requirement that it maps the positive unit section of $\tilde{i}^{*}\eta$ (which exists due to the orientation) into its analogon of $\tilde{j}^{*}\eta$. Call such an orthogonal factor to be positive.

If M is simply connected, then given f, a map $g \in C^{\infty}(M,SO(n))$ determines a differential of an immersion iff $\delta(g \cdot di \cdot f) = o$, where δ is the exterior derivative. Thus we state:

Theorem 1 Let $i \in I(M,\mathbb{R}^n)$ be fixed, then for an immersion j in the connected component O_i of i there is $g \in C^\infty(M,SO(n))$ and a bundle map $f : TM \longrightarrow TM$, positive definite and symmetric with respect to $m(i)$, such that

$$dj = g \cdot di \cdot f \quad .$$

Any Riemannian metric G on M can be represented as

$$G(X,Y) = m(i)(f^2 X,Y)$$

for any pair of smooth vector fields X,Y on M and a well defined bundle isomorphism f of TM, positive definite and symmetric with respect to $m(i)$. In general $\delta(di \cdot f) \neq o$. Call $g \in C^\infty(M,SO(n))$ an integrating factor for $di \cdot f$ if

$$\delta(g \cdot di \cdot f) = o \quad .$$

Nash's theorem [14] immediately yields:

Corollary For $n > \frac{m}{2}(3m+11)$ any smooth bundle map f of TM, positive definite and symmetric with respect to $m(i)$, admits an integrating factor $g \in C^\infty(M,SO(n))$ and

$$g \cdot di \cdot f$$

is a differential of an immersion.

Theorem 1 suggests the following interpretation for embeddings i and j: The form $di \cdot f$ reparametrizes the tangent spaces of $i(M)$ in $\mathbb{R}^n$, however, in order to envelope the integral manifold $j(M)$, they have to be displayed orthogonally by the integrating factor g to the appropriate places.

2) Connections, Torsions and Second Fundamental Tensors

Let α be a smooth $\mathbb{R}^n$-valued one form of maximal rank everywhere, regarded as a section α of $L(TM, M\times\mathbb{R}^n)$. Let $P(\alpha)$ be the unique smooth section of $L(M\times\mathbb{R}^n, TM)$ such that $P(\alpha).\alpha = id_{TM}$ (pointwise formed composition) and for which at each $p \in M$ the kernel $\ker_p P(\alpha)$ is the orthogonal complement $\alpha(T_pM)^{\perp}$ in $\mathbb{R}^n$. The form α defines a covariant derivative $\nabla(\alpha)$ by setting

$$\nabla(\alpha)_X Y = P(\alpha)\cdot d(\alpha Y)(X)$$

for all smooth vector fields X, Y on M. If α is one of the forms di or $g\cdot di$ with $g \in C^{\infty}(M, SO(n))$ the associated covariant derivative is denoted by $\nabla(i)$ or $\nabla(g)$ respectively.

Given $i, j \in E(M, \mathbb{R}^n)$ the Levi-Cività connection $\nabla(j)$ of $m(j)$, i.e. the connection associated to the form $dj = g\cdot di\cdot f$, is expressed by

$$\nabla(j)_X Y = f^{-1}\nabla(i)_X fY + f^{-1}\cdot P(i)\cdot g^{-1}\cdot dg(X)\cdot di\cdot fY =$$
$$= f^{-1}\nabla(g)_X fY .$$

The connection $\nabla(f)$ given by $f^{-1}\nabla(i)f$ leaves $m(j)$

parallel but has a non-vanishing torsion tensor $T(f)$ determined by

$$T(f)(X,Y) = f^{-1}(\nabla(i)_X(f)\cdot Y - \nabla(i)_Y(f)\cdot X) \quad .$$

Hence $f^{-1}\cdot P(i)\cdot g^{-1}\cdot dg(\ )\cdot di\cdot f$ kills this torsion. On the other hand $\nabla(g)$ leaves $m(i)$ parallel and has to be gauged by f in order to become the Levi-Cività connection of $m(j)$.

Denote by $\lrcorner$ the interior product. Observe that

$$X \lrcorner T(f) = f^{-1}\cdot P(i)\cdot (X \lrcorner \delta(di\cdot f))$$

for all smooth vector fields X on M .

Associated with the form α of maximal rank everywhere define $\tilde{\alpha} : M \longrightarrow G(m,n)$ by $\tilde{\alpha}(p) = \alpha(T_pM)$ for all $p \in M$ and let $\pi(\alpha)^{\perp} : M \longrightarrow L(M\times \mathbb{R}^n , \tilde{\alpha}^*\eta)$ be the fibrewise formed orthogonal projection from $M \times \mathbb{R}^n$ to $\tilde{\alpha}^*\eta$. The "second fundamental tensor" $S(\alpha)$ is given by

$$S(\alpha)\ (X,Y) = \pi(\alpha)^{\perp}\ d(\alpha Y)(X) \quad .$$

If $\alpha = g\cdot di\cdot f$ with $g \in C^{\infty}(M,SO(n))$ and f a smooth bundle isomorphism of TM replace $\nabla(\alpha)$ and $S(\alpha)$ by $\nabla(g,f)$ and $S(g,f)$ respectively. Then:

<u>Theorem 2</u> Let M be simply connected and $i \in I(M,\mathbb{R}^n)$ be fixed. Given $g \in C^{\infty}(M,SO(n))$ and a bundle isomorphism f of TM , symmetric and positive definite with respect to $m(i)$, the following statements are equivalent:

1) $\delta(g\cdot di\cdot f) = o$

2) $g\cdot di\cdot f = dj$ for some $j \in I(M,\mathbb{R}^n)$

3) $\nabla(g,f)$ is torsion free and $S(g,f)$ is symmetric.

In conclusion we remark that the curvature tensor $R(j)$ of $m(j)$ is

$$R(j)(X,Y)Z = f^{-1}R(g)(X,Y)fZ \ ,$$

where $R(g)$ is the curvature tensor of $\nabla(g)$.

3) The Dirac Operator

We follow here the presentation of R.Pferschy in [13] , based on [1] and [7] . Let $F(i)$ be a spin structure on the Riemannian manifold $(M,m(i))$. Call $V(i)$ its associated spinor bundle and

$$\beta(i) : \Gamma(\otimes TM) \times \Gamma(V(i)) \longrightarrow \Gamma(V(i))$$

its Clifford multiplication of tensors with spinors. Let $j \in I(M,\mathbb{R}^n)$ and $dj = g\cdot di\cdot f$. The metric $m(j)$ allows a spin structure $F(j)$ on $(M,m(j))$ which is isomorphic as a principal bundle to $F(i)$. Let $V(j)$ be the spinor bundle associated to $F(j)$ and $\beta(j)$ the corresponding Clifford multiplication. Then we have a bundle isomorphism $f_V : V(j) \rightarrow V(i)$ between the associated spinor bundles. As $F(i)$ and $F(j)$ are spin structures, the tangent bundle is associated to $F(i)$ and $F(j)$. Hence connections on M induce connections in the associated spinor bundles. Let $\nabla_V(g)$

and $\nabla_V(i)$ be the connections in $V(i)$ induced by $\nabla(g)$ and the Levi-Civitá connection $\nabla(i)$ respectively. The connection in $V(j)$ corresponding to $\nabla(j)$ is $\nabla_V(j) = f_V^{-1}\nabla_V(g)f_V$. In order to represent the Dirac operator for $m(j)$ in dependence of f and g we define the following two operators:

$$\hat{\nabla}_V(g) : \Gamma(V(i)) \longrightarrow \Gamma(T^*M \otimes V(i))$$

and

$$\hat{D}(g) : \Gamma(V(i)) \longrightarrow \Gamma(V(i))$$

by setting

$$\hat{\nabla}_V(g)_X\psi := \nabla_V(g)_{fX}\psi$$ for all smooth spinor fields ψ and vector fields X on M

and

$$\hat{D}(g) := \beta(i)\cdot\hat{\nabla}_V(g)$$ respectively.

<u>Theorem 3</u> Given the metric $m(j)$, define its Dirac operator $D(j)$ on $\Gamma(V(j))$ by $D(j) = \beta(j)\cdot\nabla_V(j)$. Then

$$D(j) = f_V^{-1}\,\hat{D}(g)f_V \quad .$$

The square of the Dirac operator is given by

$$D^2(j) = f_V^{-1}[-\,\mathrm{tr}\,\hat{\nabla}_V^{\,2}(g) + \beta(i)\cdot\hat{R}_V(g)]f_V \quad ,$$

where $\hat{R}_V(g)(X,Y)\psi := R_V(g)(fX,fY)\psi$ for all smooth vector fields X and Y and spinor fields ψ on M , respectively. $R_V(g)$ denotes the curvature tensor of $\nabla_V(g)$. On the other hand we have:

$$D^2(j) = -\,\mathrm{tr}\,\nabla_V^{\,2}(j) + \beta(j)\cdot R_V(j) \quad .$$

Now $\beta(j)\cdot R_V(j) = f_V^{-1}(\beta(i)\cdot\hat{R}_V(g))f_V$ yields:

Corollary The Laplace-Beltrami operator $\Delta_V(j)$ acting on sections ψ of $V(j)$ by $\Delta_V(j)(\psi) = - \operatorname{tr} \nabla_V^{\,2}(j)\psi$ satisfies

$$\Delta_V(j) = - f_V^{-1}(\operatorname{tr} \hat{\nabla}_V^{\,2}(g)) f_V \ .$$

4) The Equation of motion for a Perfect Fluid, which forms a Hypersurface in $\mathbb{R}^n$

Consider a film of a perfect fluid forming for each time t in an open interval $I \subset \mathbb{R}$ a hypersurface $N(t) \subset \mathbb{R}^n$, which is diffeomorphic to M. Let M be oriented. Thus the configuration space is the connected component O_i of a, say initial embedding $i \in E(M,\mathbb{R}^n)$. Clearly O_i is an open subset of the Fréchet space $C^\infty(M, \mathbb{R}^n)$, carrying Whitney's C^∞-topology.

The density of the fluid is given by a smooth function

$$\rho : O_i \longrightarrow C^\infty(M,\mathbb{R}) \quad ,$$

assigning to each $j \in O_i$ a strictly positive real-valued smooth function of M. The total mass of the fluid at the "position" j is then

$$m(j) = \int_M \rho(j)\, d\mu(j) \quad ,$$

where $\mu(j)$ is the Riemannian volume of the metric $m(j)$. The incompressibility of the fluid is given for all $h \in C^\infty(M,\mathbb{R}^n)$ by the continuity equation

$$D\rho(j)(h) + \frac{1}{2}\rho(j)\cdot tr_{m(j)} Dm(j)(h) = o \quad ,$$

where $tr_{m(j)}$ denotes the trace of the smooth two-tensor $Dm(j)(h)$ with respect to $m(j)$ (see[2]). Let us determine this density function. The decomposition for the differential of $j \in O_i$ reads as

$$dj = g\cdot di\cdot f$$

with g and f as positively orthogonal and symmetric factors respectively. Both factors depend smoothly on j. Thus a tangent vector $h \in C^{\infty}(M,\mathbb{R}^n)$ to j has as differential

$$dh = g\cdot c\cdot di\cdot f + g\cdot di\cdot k$$

for some $c \in C^{\infty}(M,so(n))$ and some bundle map k of TM, symmetric with respect to $m(i)$. Inserting the above expression into the right hand side of the continuity equation yields immediately

$$D \log \rho(j)(h) = - tr\, f^{-1}\cdot k$$

or

$$D \log \rho(j)(h) = D(\log \det f^{-1})(k) \quad .$$

Hence for each $j \in O_i$

$$\rho(j) = \rho(i)\cdot \det f^{-1} \quad .$$

Thus the mass $m(j)$ is given by

$$m(j) = \int_M \rho(i)\, d\mu(i)$$

which is certainly constant in j. Let us relate $\rho(j)$ and the mean curvature $H(j)$ of $m(j)$. Recall that

$$H(j) = - \frac{1}{2m} tr_{m(j)} Dm(j)(N(j)) \quad ,$$

where $N(j)$ is the positive unit normal section along j and $m = \dim M$. Thus for all $j \in O_i$

$$D \log \rho(j)(N(j)) = m \cdot H(j) .$$

Since, due to the symmetry of the second fundamental form of j,

$$dN(j) = - dj \cdot w_j$$

the symmetric factor $-w_j : TM \longrightarrow TM$ is the Weingarten map. The equation $N(j) = g \cdot N(i)$ implies

$$g \cdot di \cdot f \cdot w_j = - dg(\ ) \cdot N(i) + g \cdot di \cdot w_i .$$

Thus for each $j \in O_i$ the Weingarten maps of i and j are related by

$$w_j = f^{-1} \cdot w_i - f^{-1} \cdot P(i) \cdot g^{-1} \cdot dg(\ ) \cdot N(i) .$$

This equation yields

$$H(j) = \frac{1}{m} (\operatorname{tr} f^{-1} \cdot w_i - \operatorname{tr} f^{-1} \cdot P(i) \cdot g^{-1} \cdot dg(\ ) \cdot N(i)) .$$

On the other hand:

$$D \log \rho(j)(N(j)) = m \cdot H(j) .$$

The equation of motion is based on the kinetic energy $E(j)(h)$ of the fluid at each position j and speed $h \in C^\infty(M, SO(n))$:

$$E(j)(h) = \frac{1}{2} \int_M \langle h,h \rangle \rho(j) d\mu(j) = \frac{1}{2} \int_M \langle h,h \rangle \rho(i) d\mu(i) .$$

Polarization yields the weak Riemannian metric

$$G(j)(h,k) = \int_M \langle h,k \rangle \rho(i) d\mu(i) .$$

Observe that $G(j)$ is constant in j. It is now easy to see directly that the geodesics of G are straight line segments on O_i (see[2]). Compare [6] for the case that the fluid moves on a prescribed manifold.

References:

[1] Atiyah,M.F.,: Hitchin,N.J., Singer,I.M. Self-Duality in Four-Dimensional Riemannian Geometry.Proc.Royal Soc. London Ser.A 362, (1978) pp.425-461.

[2] Binz,E.: Two Natural Metrics and their Covariant Derivatives on a Manifold of Embeddings. Monatshefte für Math.89, (1980) pp.275-288.

[3] Binz,E.: The Notion of Torsion and Second Fundamental Tensor Revisited. C.R.Math.Rep.Acad.Sci.Canada, Vol.III, (1981) No.2, pp.93-98.

[4] Binz,E.: On the Levi-Cività Connection of a Gauged Levi-Cività Connection. C.R.Math.Rep.Acad.Sci.Canada, Vol.IV, (1982) No.2, pp.117-122.

[5] Binz,E.: Einstein's Evolution Equation for the Vacuum Formulated on a Space of Differentials of Immersions. Proc. of the Conference of Diff. Geom.Methods of Math.and Phys., Clausthal (1981) to appear in Lecture Notes in Math.,Springer.

[6] Ebin,D.G.,: Marsden,J.E. Groups of Diffeomorphisms and the Motion of an Incompressible Fluid. Ann.Math.92, (1970) pp.102-163.

[7] Fischer,H.R.: Dirac-Operatoren. Preprint, Amherst,Mass.,USA, (1982).

[8] Greub,W.,: Halperin,S., Vanstone,J. Connections, Curvature and Cohomology, Vol.I,II,Acad.Press, New York, (1972/73).

[9] Gutknecht,J.: Die C_{Γ}^{∞}-Struktur auf der Diffeomorphismengruppe einer kompakten Mannigfaltigkeit. Diss. ETH 5879, Zürich, (1977).

[10] Keller, H.H.: Differential Calculus in Locally Convex Spaces. Lecture Notes in Math.417, Springer,Berlin,(1974).

[11] Klingenberg,W.: Riemannian Geometry. De Gruyter Studies in Math., Berlin,New York, (1982).

[12] Peter,Th.: Zur Struktur der Immersionen einer Mannigfaltigkeit in einen Euklidischen Raum. Diss.to be submitted to University of Zurich.

[13] Pferschy,R.: Die Abhängigkeit des Dirac-Operators von der Riemannschen Metrik. Diss.to be submitted to Techn. University Graz.

[14] Sternberg,S.: Celestial Mechanics.Part II, Benjamin Inc.,New York, (1969).

LIE ALGEBRAS OF SYMMETRIES OF PARTIAL DIFFERENTIAL EQUATIONS

Yvette Kosmann-Schwarzbach

UER de Mathématiques
Université des Sciences et Techniques de Lille
59655 Villeneuve d'Ascq Cedex, France

Abstract.
The symmetries of a system of partial differential equations form a Lie algebra under the vertical bracket. Examples of Lie algebras of classical, generalized and nonlocal symmetries for diverse ordinary and partial differential equations are given. The theory of symmetries is applied to some initial-value problems and is related to the linearization problem.

Contents.

S. Sternberg (ed.), Differential Geometric Methods in Mathematical Physics, 241–277.

The problem of the determination of the symmetries of a system of partial differential equations that was solved by Sophus Lie in the 1870's [40] is now recognized as being particularly significant. Several generalizations of Lie's notion of symmetry have been introduced: generalized infinitesimal symmetries and nonlocal symmetries. They lead to the determination of new conservation laws for equations deriving from a Lagrangian, and new conserved quantities for Hamiltonian systems. In the first part of this lecture we survey the definitions and the properties of the Lie algebras of symmetries, and we outline an algorithmic procedure for the determination of these symmetries. In the second part we illustrate this theory. One type of problem consists of determining a Lie algebra of symmetries (classical, generalized or nonlocal) of a given system of partial differential equations, and of studying its structure. We illustrate this aspect of the theory in 2.1, 2.2 and 2.3. Another type of problem consists of determining which equations are invariant under a given algebra of symmetries ; the conformal Lie algebra is a popular (and important) case ; we survey this problem in 2.4. Finally, in 2.5, we mention a possible application of the theory of symmetries to the solution of some initial-value problems for partial differential equations of a very special type and in 2.6 we outline the relationship between the theory of symmetries and the linearization of partial differential equations.

1. Definitions and properties of the Lie algebra of symmetries of a system of partial differential equations.

1.1. Symmetries, generalized symmetries and nonlocal symmetries

Generalized symmetries appear in the theory of higher-order Korteweg-de Vries equations. They can be traced back to the work of Bessel-Hagen [5], to articles by R.L. Anderson, S. Kumei and C.E. Wulfman [2] and have been used by Kumei [35], [36] by Anderson and Ibragimov [1], by Olver [52] [53] [4], by Fokas [13] [17] [48] and many others for the determination of the higher-order conservation laws associated with partial differential equations derivable from variational principles. Generalized symmetries are examples of k-vector fields in the sense of R. Hermann [23] [24] and H.H. Johnson [26] [27]. We refer to [54] for an exposition of the classical theory, with numerous examples. We refer to Magri [42], Olver [52], Fokas [13] [14] and our own paper [31] for general introductions to the theory of generalized symmetries. (They have been called Lie-Bäcklund transformations in [1] and subsequent articles, although Bäcklund transformations are by no means particular cases of generalized symmetries. See 1.4.) We refer to [32] and to [28] for a detailed geometric study of generalized vector fields on fibered manifolds. We refer to Vinogradov [64][65] and to Kupershmidt [37] for an exposition of the properties of generalized vector fields considered as vector fields on the infinite jet bundle of a fibered manifold.

Generalized infinitesimal symmetries can be defined for differential operators D from a fibered manifold $\pi : F \to M$ to a fibered manifold $\pi' : F' \to M$. For simplicity we assume that F and F' are <u>vector bundles</u>. In common applications, F and F' are usually trivial, <u>i.e.</u>, $\mathbb{R}\times\mathbb{R}$ or $\mathbb{R}^2\times\mathbb{R}$.

Infinitesimal symmetries can be regarded either as projectable vector fields on F (<u>i.e.</u>, infinitesimal automorphisms of the fibered manifold F), or as first-order differential operators on the sections of F, the Lie derivation whose definition we recall below. In the same way, generalized infinitesimal symmetries can be regarded either as vertical generalized vector fields on F or as differential operators on the sections of F. Nonlocal infinitesimal symmetries are, by definition, integro-differential operators on the sections of F, but their interpretation as geometric objects on F is not yet clear, except in the case where F is itself a jet bundle of a vector bundle, F_0. Then integro-differential operators on the sections of F correspond to differential operators on the sections of F_0 and this reduces to the preceding case, generalized infinitesimal symmetries. We infer from [58] and from the reference it makes to [66], a book by Vinogradov that I have not yet been able to consult, that Vinogradov has initiated a theory of nonlocal symmetries which takes into account the constants of integration that can be added arbitrarily to each integral.

Any <u>projectable vector field</u> X on $\pi : F \to M$ defines a differential operator on the sections of F, called the <u>Lie derivation</u> with respect to X and denoted, again, by X. If μ_t is the flow of X, and if u is a section of F, then, by definition, $Xu = \frac{d}{dt}(\mu_t \cdot u)\Big|_{t=0}$ where for each $x \in M$, $(\mu_t \cdot u)(x) = \mu_t(u(\mu_{tM}^{-1}x))$, μ_{tM} being the flow of the projection X_M of X onto M. In local coordinates $x = (x^i)$ on the base, and $y = (y^\alpha)$ in the fiber, if $X(x,y) = X^i(x)\partial_i + Y^\alpha(x,y)\partial_\alpha$, then $(Xu)(x) = Y^\alpha(x,u(x)) - X^i(x)u_i(x)$, where subscripts denote derivatives.

More generally, a <u>vertical generalized vector field</u> of order k on F is a mapping $X : J^kF \to VF$, such that $X(j_x^k u) \in V_{u(x)}F$ for each k-jet $j_x^k u$ at $x \in M$, where VF denotes the vertical tangent bundle of F. Such a mapping also defines a <u>differential operator</u> on the sections of F defined by

$$(Xu)(x) = X(x,u(x),u_i(x),\dots,u_{I(k)}(x)),$$

where $I(k)$ stands for all multiindices of length k. We say that the vertical generalized vector field X has a flow through u if the partial differential equation of evolution $U_t = XU$ has a unique solution $U(t,x)$ satisfying the initial condition $U(0,\cdot) = u$, for t in some interval containing 0, and we set $U(t,x) = (\mu_t u)(x)$. Thus by definition the flow μ_t of X satisfies $\frac{\partial}{\partial t}(\mu_t u) = X(\mu_t u)$.

<u>Fréchet derivative of an operator</u>. If D is a differential operator from F to F', we define the operator VD from $F \times F$ to F',

$$VD(u,v) = \frac{d}{dt} Du^{(t)}\Big|_{t=0} ,$$

where $u^{(t)}$ is a 1-parameter family of sections of F such that $u^{(0)} = u$ and $\frac{du^{(t)}}{dt}\Big|_{t=0} = v$. VD is linear in the second variable, v. VD is known in the literature by different names : Fréchet derivative, variation, linearized operator or vertical operator of D.

Notation. For X a vertical generalized vector field and u a section of F, we set $(VD\circ X)(u) = VD(u,Xu)$.

Lemma. In local coordinates, for D, a differential operator of order r,

$$VD(u,v) = \frac{\partial D}{\partial u^\alpha} v^\alpha + \frac{\partial D}{\partial u_i^\alpha} v_i^\alpha + \ldots + \frac{\partial D}{\partial u_{I(r)}^\alpha} v_{I(r)}^\alpha .$$

Example. Let $F = F' = \mathbb{R}\times\mathbb{R} \to \mathbb{R}$. Let $Du = u_{xx}+uu_x$. Then $VD(u,v) = v_{xx}+uv_x+u_xv$.

Let $Xu = xu_xu_{xx}-u_x^2$. Then

$$(VD\circ X)(u) = (xu_xu_{xxxx}+3xu_{xx}u_{xxx}) + u(xu_xu_{xxx}+xu_{xx}^2-u_xu_{xx})$$
$$+ u_x(xu_xu_{xx}-u_x^2).$$

Symmetries. Let X be a projectable vector field on F. We say that X is an infinitesimal symmetry of D if there exists a projectable vector field X' on F' such that $X'(0) = 0$ and

$$VD \circ X - X' \circ D = 0.$$

This condition implies that $D^{(t)} = D$, where $D^{(t)}$ is obtained from D by the action of the flows μ_t of X and μ_t' of X'. In particular, this condition further implies that the set of solutions of D is infinitesimally invariant under X.

More generally,

<u>Definition</u>. Let X be a vertical generalized vector field on F. We say that X is a <u>generalized infinitesimal symmetry</u> of D if there exists a differential operator X' from $F\times F'$ to F' such that $X'(u,0) = 0$ for each section u of F and

$$VD \circ X - X' \circ D = 0, \tag{1}$$

<u>i.e.</u>, $VD(u,Xu) - X'(u,Du) = 0$ identically in u.

When X' is a polynomial in v (and its derivatives) it can be assumed, without loss of generality, to be linear. In fact we can replace $X'(u,v)$ by any operator $X''(u,v)$ such that $X'(u,Du) = X''(u,Du)$ identically. If, for example, $X'(u,v) = uu_x e^u v^4 v_x^2$, we can replace X' by $X''(u,v) = uu_x e^u (Du)^3 (Du)_x^2 v$, and similarly for any monomial in $v^{\alpha'}, v_i^{\alpha'}, \ldots, v_{I(k)}^{\alpha'}$.

From this definition we see that $(VD \circ X)(u) = 0$ whenever Du and the successive derivatives of Du are zero. Thus the essential property of the symmetries is preserved in this generalization : "$Du = 0$ implies that $(VD \circ X)(u) = 0$" is the infinitesimal version of "$Du = 0$ implies that $D(\mu_t u) = 0$" where μ_t is the flow of X. It must be noted, however, that the flow may not exist in the case of a generalized vector field. In [64] [65] this infinitesimal property is expressed geometrically by the condition that X be tangent to the infinite prolongation of the equation $Du = 0$. Here, in order to facilitate the computations, we impose the slightly stronger condition that $(VD \circ X)(u)$ be, for each u, a function of Du and its successive derivatives. This last condition is identical to Olver's nondegeneracy condition [52].

It follows from the definition that, if X is a generalized infinitesimal symmetry of D whose flow μ_t is defined for some value t, for each solution u of $Du = 0$, the image $\mu_t u$ of u under the flow of X is also a solution of D.

This fundamental property is the basis of many of the applications of the theory of symmetries. (See 1.3 and 2.5.)

We shall sometimes write symmetry instead of infinitesimal symmetry and generalized symmetry instead of generalized infinitesimal symmetry. To distinguish the symmetries from the generalized symmetries we shall sometimes refer to the former as classical symmetries.

Nonlocal symmetries. More generally, let X be an integro-differential operator on the sections of F which incorporates at least one integration. If there exists an integro-differential operator X' on the sections of $F \times F'$ such that $X'(u,0) = 0$ for each section u of F, and $VD \circ X - X' \circ D = 0$ (still carrying over our stronger assumption), we say that X is a nonlocal generalized infinitesimal symmetry of D or, for short, a nonlocal symmetry of D.

For example, if one considers real-valued, rapidly decreasing functions of one variable, X may be a function of u, of its derivatives, of $\int u$ where $(\int u)(x) = \int_{-\infty}^{x} u(\xi)d\xi$, and of successive integrals, and $(VD \circ X)u = 0$ may be a consequence of $Du = 0$ and its integro-differential consequences.

In [29] Konopelchenko and Mokhnachev extend to this case the method of prolongation of Anderson and Ibragimov [1]. They also generalize several standard results from the local case to the nonlocal one. Among other results, they prove that the symmetries, including the nonlocal ones, of the integral consequences of an equation are the same as those of the equation itself, and they obtain the conservation laws associated with equations deriving from a Lagrangian invariant under a nonlocal symmetry.

1.2. Vertical bracket.

What we call the vertical bracket of vertical generalized vector fields appears in the work of Segal [60] and of Lax [38]. It plays a fundamental role in the geometric theory of symmetries as it is expounded by Vinogradov [64][65] and Kupershmidt [37], in Manin [46], in the papers of Olver [49][53] and in our own articles [32][33]. Vinogradov calls it the Jacobi bracket while the others do not designate it by a special term.

Let X_1 and X_2 be two vertical generalized vector fields on F, i.e., differential operators on the sections of F. The vertical bracket $[X_1,X_2]_V$ of X_1 and X_2 is the vertical generalized vector field defined by

$$[X_1,X_2]_V = VX_1 \circ X_2 - VX_2 \circ X_1.$$

If we associate with a vertical generalized vector field X its infinite prolongation $\mathrm{pr}X$, we obtain a vector field on the infinite jet bundle of F. The vertical bracket is the opposite of the Lie bracket of vector fields on the infinite jet bundle of F, more precisely,

$$\mathrm{pr}([X_1,X_2]_V) = -[\mathrm{pr}X_1,\mathrm{pr}X_2].$$

If X_1 and X_2 are ordinary, projectable vector fields on F, their vertical bracket is the opposite of the usual Lie bracket of vector fields.

If X_1 and X_2 are linear differential operators on the sections of F, they coincide with their linearized operators, and their vertical bracket is the usual commutator of linear differential operators. Therefore, the vertical bracket is the natural generalization to nonlinear operators of the commutator of two linear operators. It coincides with the commutator of nonlinear operators used extensively by Magri [41][42].

Finally, if X_1 and X_2 are polynomial differential operators on a trivial vector bundle with base $\mathbb{R}$, their vertical bracket is the opposite of the Gelfand-Dikii bracket [20].

<u>Proposition</u>. (See <u>e.g.</u> [32] 1.24). The vertical bracket is a Lie algebra bracket.

The notion of Fréchet derivatives, and therefore that of vertical brackets, extends without difficulty to integro-differential operators at least for operators on functions of one independent variable, provided the integrals exist. We remark that if one considers the subspace of the integro-differential operators depending on u, its derivatives and its first-order integrals, it is not a Lie subalgebra of the Lie algebra of all nonlocal symmetries under the vertical bracket. In fact, the bracket of two operators within that subspace in general involves higher-order integrals of some functions of u and its derivatives, as the following example illustrates.

<u>Example</u>. Let u be a real-valued, rapidly decreasing function of one variable, x. Let $Xu = u\int u$ and $Yu = u_x^2\int u$. Then $VX(u,v) = u\int v + v\int u$ and $VY(u,v) = u_x^2\int v + 2u_x v_x\int u$. Therefore

$$[X,Y]_V(u) = u\int(u_x^2\int u) + u_x^2(\int u)^2 - 2u_x(u_x\int u + u^2) - u_x^2\int(u\int u)$$

$$= u\int(u_x^2\int u) - 2u_x^2\int u + \frac{1}{2}\,u_x^2(\int u)^2 - 2u_x u^2,$$

provided that the integrals exist. Note that $\int(u_x^2\int u)$ cannot be reduced to a first-order integral of u.

In [67], dealing with evolution partial differential equations $u_t = Xu$ on functions u of two variables, t and x, Vinogradov and Krasil'ščik introduce nonlocal symmetries depending upon a new variable p such that $p_x = u$ and $p_t = Xu$, <u>i.e.</u>, $p = \int u$ + constant, for u a solution of the equation. These are in fact families of nonlocal symmetries depending upon an arbitrary constant of integration. They stress the fact that in this theory the bracket of two nonlocal symmetries is not uniquely defined. This is not surprising since arbitrary constants of integration, which can be functions of t, enter the brackets non trivially. We shall consider their example in 2.3.

1.3. The case of evolution equations.

We now consider evolution equations in p space variables and one time variable t. To this end we introduce the projection ρ of $M\times\mathbb{R}$ onto M and the induced vector bundle ρ^*F over $M\times\mathbb{R}$. An evolution equation can be written

$$U_t = PU,$$

where P is a differential operator on ρ^*F that involves only spatial derivatives, and U is a section of ρ^*F. We say that P is time-independent if it is induced from a differential operator on F.

Proposition [53]. Let Q be a vertical generalized vector field on ρ^*F. Q is a symmetry of $D = -\frac{\partial}{\partial t} + P$ if and only if

$$[P,Q]_V = Q_t. \tag{2}$$

For a time-independent Q to be a symmetry of P it is necessary and sufficient that $[P,Q]_V = 0$. This condition expresses the fact that the flows of P and Q (if they exist) commute. (See [31].)

We note that, for evolution equations, the invariance condition is notably simpler than for general equations since condition (1) is satisfied if and only if $VD \circ Q - VQ \circ D = 0$ (necessarily $Q' = VQ$).

This property of the symmetries of evolution equations is the basis of the article [25] where Ibragimov and Shabat prove that any Lie algebra of generalized symmetries of a polynomial partial differential equation of evolution in the space variable x whose operator P is equal to $\frac{\partial^k}{\partial x^k}$ plus terms of lower order is Abelian.

1.4. <u>Algorithms for the determination of generalized infinitesimal symmetries and of Bäcklund transformations between evolution equations.</u>

The most efficient method for determining a sequence of higher-order symmetries of a given system of partial differential equations is the discovery of a <u>recursion operator</u> in the sense of Olver [49]. See [13][16][17]. This method does not in general yield all the symmetries, so the cumbersome, direct methods which we consider in this paragraph retain their value. Other methods which take into account a priori considerations of polynomial degree and "grading" have been successfully applied in [67].

For a given polynomial differential operator D acting on d scalar functions of n independent variables, the problem of determining all the polynomial symmetries of D of maximal degree q (as a polynomial in the partial derivatives of the unknown functions) and maximal order k (the highest order of the partial derivatives) can be solved by an algorithm. The polynomial form of the symmetries X is an assumption. It would be interesting to prove that this assumption does not entail a loss of generality when D is polynomial. To determine the polynomial symmetries of D we write that the relation (1) is satisfied identically, equating the coefficient of each monomial to 0. The unknowns are the coefficients of X and of X'. Another method (which, unlike the first, determines only X and not X') consists of writing $(VD \circ X)(u)$ or $[D,X]_V(u) = (VD \circ X - VX \circ D)(u)$ and expressing that **either is** a linear combination (whose coefficients are functions of u) of Du and its derivatives. In both methods, the coefficients have to satisfy a largely overdetermined system of partial differential equations. Unless D is simple enough to admit symmetries the system's only solution is, of course the trivial one, 0. The number of coefficients of X is $K(q,K(n,k)d)d$ where $K(n,k) = 1 + n + \dots + C^k_{n+k-1}$ and this number, like the number of equations, increases very rapidly with the degree q and the order k. For this reason, the necessary calculations require a computer. A. Bomberault (Brooklyn College, C.U.N.Y.) has programmed the second algorithm using symbolic manipulation in FORMAC (unpublished). A subprogram determines the Fréchet derivative of an operator. Bomberault then introduces an ordering (called 'length') of the monomials of powers of partial derivatives such as (for two functions, u and v, of two variables, x and y) $u_x u_{xxy}^2 vv_{xy}^3 v_{yy}$. With this ordering he can assign to any system of polynomial differential equations an infinite-dimensional subspace F_D in an infinite-dimensional vector space F such that X is a symmetry of D if and only if the Fréchet derivative of D, evaluated on X, belongs to F_D. Multiplication by functions and taking partial derivatives are linear operators on the vector space F. The determination of the symmetries X of length ℓ of a differental operator D of length h reduces to the determination of those vectors X such that $VD \circ X$ belongs to a finite-dimensional subspace of F_D.

Algorithms [63] and computer programs using MACSYMA [56] and REDUCE [59] have been written for the determination of (classical) infinitesimal symmetries. REDUCE 2 has also been used in [10] [21] to find prolongation Lie algebras in the sense of Wahlquist and Estabrook, and symmetries of exterior differential systems. Also in the U.S.S.R. computer programs (CINO and PASSIV) have been written to obtain the differential equations determining the infinitesimal symmetries of a given equation. (See the references in [54] which dates from 1978.)

Algorithm for the determination of Bäcklund transformations for evolution equations. Let $U_t - PU = 0$ be an evolution equation, the number of whose unknown functions, d, and the number of whose space variables p, are arbitrary. A Bäcklund transformation between that evolution equation and another evolution equation, $\bar{U}_t - \bar{P}\bar{U} = 0$, the number of whose unknown functions is $\bar{d}$ but the number of whose space variables is p, is a bidifferential operator $B(u,\bar{u})$ on pairs of vector-valued functions of p variables such that $(P,\bar{P})$ is a generalized infinitesimal symmetry of B, and B has an injective partial Fréchet derivative with respect to $\bar{u}$ (see [31] and [16]). The condition for B to be a Bäcklund transformation can therefore be written

$$VB(u,\bar{u},Pu,\bar{P}\bar{u}) - \mathcal{P}'(u,\bar{u},B(u,\bar{u})) = 0 . \qquad (3)$$

B, $\bar{P}$ and $\mathcal{P}'$ are the unknowns. They are assumed to be polynomial differential operators in the suitable variables; B is a polynomial of (arbitrary) degree q in the partial derivatives (with respect to the space variables) of order 0 to k of the functions $(u,\bar{u}) = (u^1,\ldots,u^d,\bar{u}^1,\ldots,\bar{u}^{\bar{d}})$ whose coefficients are functions of the space variables ; $\mathcal{P}'(u,\bar{u},w)$ is linear in w, the coefficients being polynomials in the partial derivatives of order 0 to k of the functions $(u,\bar{u})$. All the integers, $\bar{d}$, the number of scalar components of B, the order of B, the order of $\bar{P}$ and $\mathcal{P}'$, the polynomial degrees of $\bar{P}$, B and $\mathcal{P}'$, must be chosen before the beginning of the computation. The left-hand side of the equation is a polynomial in the partial derivatives of order 0 to $k+r$ (where r is the higher of the orders of the two evolution equations) of the functions $(u,\bar{u})$. By setting the coefficient of every term equal to 0, we obtain a very large set of homogeneous linear partial differential equations in the unknown coefficients of B, $\bar{P}$ and $\mathcal{P}'$. Any solution of this system will furnish a Bäcklund transformation of order and of polynomial degree less than prescribed integers, of the given equation. By increasing the order and the degree, additional Bäcklund transformations will be obtained, if they exist.

In the case of auto-Bäcklund transformation, the method is somewhat more tractable because $\bar{d}$ is taken to be equal to d and $P = \bar{P}$. The number of unknowns is thus reduced but the number of conditions remains unchanged ; there will be fewer equations admitting auto-Bäcklund transformations than the more general

Bäcklund transformations, as is to be expected.

There are few definitions of Bäcklund transformations for partial differential equations which cannot be written in evolution form, and they do not lend themselves to computations of the above type.

2. Examples and applications.

2.1. Lie algebras of classical symmetries.

Burgers's equation, $u_t - uu_x - u_{xx} = 0$. The Lie algebra [54] [39] [30] of all (classical) symmetries of Burgers's equation is a 5-dimensional Lie algebra with generators

$$X_1 u = u_x \qquad X_2 u = u_t \qquad X_3 u = tu_x + 1$$

$$X_4 u = xu_x + 2tu_t + u \qquad X_5 u = txu_x + t^2 u_t - (x - tu).$$

It is the semi-direct product of the Abelian ideal generated by X_1 and X_3 with the semi-simple subalgebra generated by X_2, X_4 and X_5.

Korteweg-de Vries equation, $u_t - uu_x - u_{xxx} = 0$. The Lie algebra [34] [30] of all (classical) symmetries of the KdV equation is a 4-dimensional solvable Lie algebra with generators

$$X_1 u = u_x \qquad X_2 u = u_t \qquad X_3 u = tu_x + 1$$

$$X_4 u = xu_x + 3tu_t + 2u.$$

Many other examples are known. See, e.g., [6] [47] [51], and [54] for a wealth of examples.

2.2. Generalized symmetries of the ordinary differential equation $u_x^2 + u = 0$.

We have computed by hand the polynomial symmetries of some simple, ordinary differential equations. We state the results for the nonlinear, first-order operator $Du = u_x^2 + u$.

Among the symmetries of D are the operators $X_{(a)} = aD$ where a is an arbitrary function of x. (This is true for any operator D since $VD(u,aDu)$ is obviously 0 when $Du = 0$.) Here $[X_{(a)},D]_V = 2a_x u_x Du$.

Among the other symmetries X of order ≤ 1, we first computed those of degree ≤ 2, and for each X the corresponding X', or rather the Z satisfying $[X,D]_V = Z \circ D$.

$$X_1 u = u_x \qquad Z_1(u,v) = 0$$

$$X_2 u = xu_x - 2u \qquad Z_2(u,v) = 0$$

$$X_3 u = -\tfrac{1}{4}x^2 u_x + xu + uu_x \qquad Z_3(u,v) = -u_x v$$

$$X_4 u = \tfrac{1}{8}x^3 u_x - \tfrac{3}{4}x^2 u - \tfrac{3}{2}xuu_x + u^2 \qquad Z_4(u,v) = -\tfrac{3}{2}xu_x v - uv$$

Note that only X_1 (a translation) and X_2 (a scale transformation) are classical infinitesimal symmetries: the invariance of $Du = 0$ under X_1 and X_2 implies that if $u(x)$ is a solution of $Du = 0$, so too are $u(x-t)$ and $e^{2t}u(e^{-t}x)$.

It appears that, if we assign to x the grade 1, to u the grade 2, and therefore to u_x the grade 1, to u_{xx} the grade 0, to u_{xxx} the grade -1, etc., each X_p $(p = 1,\ldots,4)$ is of grade p. We define inductively the symmetries $X_5, X_6, \ldots$ by the recursion formula, for $p \geq 4$,

$$X_{p+1} = [X_3, X_p]_V ,$$

so that $Ad_V X_3$ is a recursion operator generating the sequence of symmetries $X_5, X_6, \ldots, X_p, \ldots$ Each X_p, for $p \geq 5$, is a symmetry of grade p whose highest term in x is $(-1)^p \frac{(p-4)!}{2^{2p-5}} x^{p-1} u_x$. We find for instance that

$$X_5u = -\frac{1}{32}x^4u_x + \frac{1}{4}x^3u + \frac{3}{4}x^2uu_x - xu^2 - \frac{1}{2}u^2u_x$$

$$X_6u = \frac{1}{64}x^5u_x - \frac{5}{32}x^4u - \frac{5}{8}x^3uu_x + \frac{5}{4}x^2u^2 + \frac{5}{4}xu^2u_x - \frac{1}{2}u^3 .$$

Here is the multiplication table for the brackets $[X_p,X_q]_V$, for $1 \leq p \leq 3$ and $1 \leq q \leq 6$. (We have omitted the subscript V.)

$[X_1,X_2] = X_1$	$[X_1,X_3] = -\frac{1}{2}X_2$	$[X_1,X_4] = -\frac{3}{2}X_3$
$[X_2,X_3] = X_3$	$[X_2,X_4] = 2X_4$	
$[X_3,X_4] = X_5$	$[X_3,X_5] = X_6$	$[X_3,X_6] = X_7$
$[X_1,X_5] = -X_4$	$[X_1,X_6] = -\frac{5}{2}X_5$	
$[X_2,X_5] = 3X_5$	$[X_2,X_6] = 4X_6$.	

More generally, the brackets verify $[X_p,X_q]_V = \mu_{p,q}X_{p+q-2}$ for $p \geq 1$ and $q \geq 1$. The values of the rational numbers $\mu_{p,q}$ for small p and q are given above. The following relations hold ((ii) and (iii) determine $\mu_{p,q}$ by recursion):

i) for $p \geq 5$, $\mu_{3,p} = 1$, $\mu_{2,p} = p-2$, $\mu_{1,p} = -\frac{(p-1)(p-4)}{4}$.

ii) for $p \geq 5$, $\mu_{4,p} = \frac{6+(p-1)(p-4)\mu_{4,p-1}}{(p+1)(p-2)}$.

iii) for $p \geq 5$, $q \geq 5$, $\mu_{p,q} = \frac{(p-1)(p-4)\mu_{p-1,q}+(q-1)(q-4)\mu_{p,q-1}}{(p+q-3)(p+q-6)}$

We also note that for $p \geq 4$, $\mu_{p,p+2} = \mu_{p,p+1}$ and for $p \geq 4$, $q \geq 4$,

$$\mu_{p+1,q} = \mu_{p,q} - \mu_{p,q+1} .$$

Thus

$$\mu_{4,5} = \mu_{4,6} = 1/3, \quad \mu_{4,7} = 3/10, \quad \mu_{4,8} = 4/15, \quad \mu_{4,9} = 5/21, ..$$

$$\mu_{5,6} = \mu_{5,7} = 1/10, \quad \mu_{5,8} = 1/35, ...$$

Moreover, for $p \geq 2$, $\frac{\partial}{\partial x} X_p = [X_1, X_p]_V = \mu_{1,p} X_{p-1}$.

The symmetries X_p, for $p \geq 1$, generate a Lie algebra of symmetries of order ≤ 1 of D. This Lie algebra possesses four finite-dimensional subalgebras generated by the following pairs of symmetries : X_1, X_2 (the ordinary symmetries), X_2, X_3 , X_2, X_4 and X_1, X_3. Three infinite-dimensional subalgebras are generated by X_3, X_4, X_2, X_3, X_4 and X_1, X_2, X_4 (or X_1, X_3, X_4). The Lie algebra $\mathfrak{sl}(2,\mathbb{R})$ has a representation as a Lie algebra of symmetries of D defined by

$$\begin{pmatrix} 1 & 0 \\ 0 & -1 \end{pmatrix} \to \frac{1}{2} X_2 \quad , \quad \begin{pmatrix} 0 & 1 \\ 0 & 0 \end{pmatrix} \to X_3 \quad , \quad \begin{pmatrix} 0 & 0 \\ 1 & 0 \end{pmatrix} \to X_1 .$$

The symmetries of the form $X_{(a)}$ form a Lie algebra. The bracket of X_1 with $X_{(a)}$ is $X_{(a_x)}$; the bracket of X_2 with $X_{(a)}$ is $X_{(xa_x)}$ so that $X_{(a)}$, X_1, X_2 generate an infinite-dimensional Lie algebra of symmetries in which the symmetries of the form $X_{(a)}$ generate an ideal. The bracket of X_3 with $X_{(a)}$ (with a of grade k) is $\bar{a}D$ where $\bar{a}$ is the operator of grade k+1 defined by $\bar{a}u = -au_x + a_x u - a_x x^2$. In particular, $[X_3, D] = -u_x D$.

We can also compute the symmetries of order 2. We find $X^2_{(b)} = bD_x$ defined by $X^2_{(b)}(u) = b(Du)_x = b(u_x + 2u_x u_{xx})$, where b is an arbitrary function of x. These symmetries satisfy $[X^2_{(b)}, D]_V u = 2\, bu_x (Du)_{xx}$. Below are the other symmetries of order 2 and of degree ≤ 2 and their corresponding X':

$X_1^2 = u_x u_{xx}$ $\qquad$ $X'^2_1 = u_x v_{xx}$

$X_2^2 = xu_x u_{xx} - u_x^2$ $\qquad$ $X'^2_2 = -u_x v_x + xu_x v_{xx}$

$X_3^2 = \frac{1}{2}x^2 u_x u_{xx} - xu_x^2 + uu_x$ $\qquad$ $X'^2_3 = uv_x - xu_x v_x + \frac{1}{2}x^2 u_x v_{xx}$

$X_4^2 = -\frac{1}{4}x^3 u_x u_{xx} + \frac{3}{4}x^2 u_x^2 - \frac{3}{2}xuu_x + u^2$

$$X'^2_4 = uv - \frac{3}{2}xuv_x + \frac{3}{4}x^2 u_x v_x - \frac{1}{4}x^3 u_x v_{xx} .$$

If we search for the nonlocal symmetries of D, setting $u_{-1} = \int u$, we find that the only nonlocal symmetry of the form $Xu = au_{-1} + bu + cu_x + duu_{-1} + euu_x + fu_x u_{-1}$, is

$$Xu = -\frac{1}{12}x^3 u_x + \frac{1}{2}x^2 u + xuu_x - u_x u_{-1} .$$

2.3. Nonlocal symmetries of Burgers's equation.

In [49] Olver showed that Burgers's equation $Du \equiv u_t - uu_x - u_{xx} = 0$ possesses a recursion operator which permits the computation of an infinite sequence of generalized symmetries. The recursion operator is nonlocal but the symmetries that it generates are local. Nonlocal symmetries of the equation were determined by Vinogradov and Krasil'ščik [67]. The nonlocal symmetries of D which depend on x, t, u and $\int u$ are defined by

$X_{C,K}u = (Cu - 2C_x)e^{-\frac{1}{2}\int u + K}$ where C is a function of x and t satisfying $C_t - C_{xx} + K_t C = 0$. (In [67] the condition is written as $C_t + C_{xx} + K_t C = 0$ which seems to be a misprint.) The condition on C and K for $X_{C,K}$ to be a symmetry of D is easily found because a calculation shows that, for any functions $C(x,t)$ and $K(t)$,

$$VD(u,X_{C,K}u) = \Big(-\tfrac{1}{2}(Cu-2C_x)\int Du+CDu+(C_t-C_{xx}+K_tC)u - 2(C_t-C_{xx}+K_tC)_x\Big)e^{-\frac{1}{2}\int u+K}.$$

Setting $\tilde{C} = Ce^K$ it appears that any such symmetry can be written $X_{\tilde{C}}u = (\tilde{C}u-2\tilde{C}_x)\,e^{-\frac{1}{2}\int u}$, with $\tilde{C}_t-\tilde{C}_{xx} = 0$.

In order to compute the vertical bracket of two nonlocal symmetries, X_A and X_B, we first determine $VX_A(u,v) = (Av - \frac{1}{2}(Au-2A_x)\int v)e^{-\frac{1}{2}\int u}$. We see that only if $B_x = 0$ is it possible to take $\int X_B$ to mean (if the integral is defined) $\int_{-\infty}^{x}(Bu-2B_x)\,e^{-\frac{1}{2}\int_{-\infty}^{\xi}u}\,d\xi$. One is thus led to take $\int v$ to be defined up to an arbitrary constant, <u>i.e.</u>, a function of t. Thus $\int X_Bu = \int (Bu-2B_x)\,e^{-\frac{1}{2}\int u} = -2(Be^{-\frac{1}{2}\int u} + \varepsilon)$, ε being a function of t only, and $VX_A(u,X_Bu) = A(Bu-2B_x)e^{-\int u} + (Au-2A_x)Be^{-\int u} + (Au-2A_x)\varepsilon\, e^{-\frac{1}{2}\int u}$. Setting $\int X_Au = -2(Ae^{-\frac{1}{2}\int u} + \gamma)$, we find

$$\begin{aligned}[X_A,X_B]_V(u) &= (\varepsilon(Au-2A_x) - \gamma(Bu-2B_x))e^{-\frac{1}{2}\int u}\\ &= ((\varepsilon A-\gamma B)u - 2(\varepsilon A-\gamma B)_x)\,e^{-\frac{1}{2}\int u}\\ &= X_{\varepsilon A-\gamma B}(u).\end{aligned}$$

If $A_t-A_{xx}+K_tA = 0$ and $B_t-B_{xx}+L_tB = 0$, then $(\varepsilon A-\gamma B)_t - (\varepsilon A-\gamma B)_{xx} = (\varepsilon_t-\varepsilon K_t)A - (\gamma_t-\gamma L_t)B$. The right-hand side can be written as $M(\varepsilon A-\gamma B)$, where M is a function of t, provided that either $AB_x - A_x B = 0$ or $\varepsilon\gamma_t - \gamma\varepsilon_t = \varepsilon\gamma(L_t - K_t)$, that is, either $B = \lambda(t)A$ or $\gamma = c\varepsilon e^{L-K}$ (where c is a constant). If $B = \lambda(t)A$, then $\varepsilon A - \gamma B$ is of the form $\phi(t)A$; we can verify directly that if X_A is a symmetry, $X_{\phi(t)A}$ is a symmetry. If $\gamma = c\varepsilon e^{L-K}$, then $\varepsilon A - \gamma B = \varepsilon e^{-K}(Ae^{K} - cBe^{L})$, and $X_{\varepsilon A - \gamma B}$ is again a symmetry. In the case where $K = L = 0$, we verify directly that if X_A and X_B are symmetries satisfying $A_t - A_{xx} = B_t - B_{xx} = 0$, then $\phi(t)(A-cB)$ is a symmetry.

The X_C-invariant solution of $Du = 0$ is $u = \frac{2C_x}{C}$, where C is a solution of $C_t - C_{xx} + K_t C = 0$. This is a way of recovering the Hopf-Cole transformation.

2.4. Conformal covariance.

The determination of the conformally covariant wave equations is an important problem in physics and in mathematics since those wave operators that are conformally covariant provide intertwining operators for representations of the conformal group (or Lie algebra) of space-time. The M.I.T. school has produced important work on this problem. (See [55] and the references to earlier papers it contains.) The conformal covariance of the classical wave equations (Laplace, Maxwell and Dirac) on Minkowski space have long been investigated and the results have been generalized to arbitrary space-times and even to manifolds of arbitrary dimension. It was already known in the 1950s that in order to make the Laplacian on functions conformally covariant on curved space-time, the curvature term, $\frac{1}{6}R$, where R is the Riemannian curvature, had to be added ; generalizations of this result are now known for arbitrary pseudo-Riemannian manifolds and for Laplacians on forms, suitably modified [7]. Gürsey [22] proved that a nonlinear interaction term can be added to Dirac's equation on Minkowski space without destroying the conformal covariance. This result generalizes to arbitrary spin manifolds and to any operator possessing a 'conformal bidegree'. The exponent of the nonlinear interaction term has a simple expression in terms of the conformal bidegree of the operator [9]. We refer to this last article for more references, both historical and current. Recently, conformally covariant first-order operators on higher-order spinors [55] and on spinor-forms [8] have been discovered.

In all the preceding examples, each of the two representations of the conformal Lie algebra which the operator intertwines is the standard representation on fields of tensors or spinors multiplied by a scalar factor equal to a suitable power of the ratio of the conformal transformation. It should be clear from the rest of this lecture that other types of representations can be considered : the generators of the conformal Lie algebra of the space-time can be mapped into generalized infinitesimal symmetries, or even into nonlocal infinitesimal symmetries of the operators, although the physical interpretation of conformal covariance in this extended sense is not clear. Results along these lines are rather scarce. We shall cite [19] and the papers quoted there. In [19] Fushchich and Nikitin determine a representation of the conformal Lie algebra of Minkowski space by nonlocal symmetries of each Poincaré-invariant wave equation with zero mass and discrete spin, which reduce to local symmetries in the well-known case of Dirac's equation. Fushchich and Nikitin also prove that there exists a 23-dimensional Lie algebra of symmetries (some nonlocal) of Dirac's equation or of Maxwell's equation. It would be interesting to investigate whether it is also a Lie algebra of symmetries of some nonlinear modifications of Dirac's and Maxwell's equations and to study the symmetry-breaking interactions.

2.5. Application of the properties of symmetries to initial-value problems.

We consider a first-order partial differential operator D with a known general solution, $u = u(x,\alpha)$, where α is an arbitrary constant. Even in such a simple case, the higher-order symmetries of D can be very complicated, as the example in 2.2

shows. Let X be such a symmetry of D. We can use the fundamental property of the symmetries of D to integrate the evolution equation $U_t = XU$ whenever the initial condition, $U(0,x)$, is a solution u of D. This last requirement is very special and the method will not extend to other cases. Replacing $U(t,x)$ by the expression for the general solution of D, where α is now a function of t, we obtain an ordinary differential equation for α as a function of t (whose coefficients might depend on x). Solving for α, if possible, we obtain the solution $U(t,x)$ of the initial-value problem in question as $u(x,\alpha(t,x))$. Thus we obtain exact solutions of some very particular initial-value problems for some nonlinear, higher-order partial differential equations.

More precisely, let the initial condition be $U(0,x) = u(x,\alpha_o)$ where α_o is a constant. Replacing U by $u(x,\alpha)$ in XU, we obtain a function $f(x,\alpha)$; $U(t,x) = u(x,\alpha(t,x))$ will be a solution of $U_t = XU$ if and only if α satisfies the ordinary differential equation

$$\frac{\partial u}{\partial \alpha}(x,\alpha)\frac{d\alpha}{dt} = f(x,\alpha).$$

U will satisfy the initial condition $U(0,x) = u(x,\alpha_o)$ if and only if this equation possesses a solution α verifying $\alpha(0,x) = \alpha_o$. This equation for α is separable and can be integrated in terms of a primitive $G(x,\alpha)$ of $\frac{1}{f(x,\alpha)}\frac{\partial u}{\partial \alpha}(x,\alpha)$, which satisfies $G(x,\alpha) = t + G(x,\alpha_o)$. This relation defines α implicitly as a function of x, t and α_o. Replacing α by this function in $u(x,\alpha)$ yields the solution $U(t,x)$ (which depends on α_o). This method will, in principle, be applicable even when X is not quasilinear, although in this case $\frac{1}{f}\frac{\partial u}{\partial \alpha}$ is likely to be a complicated function of α, and it may be impossible to calculate a primitive $G(x,\alpha)$ explicitly and to invert it.

Example 1. We consider $Du = u_x^2+u$ as in 2.2. The general solution of $Du = 0$ is $u(x,\alpha) = -\frac{1}{4}(x+\alpha)^2$ where α is an arbitrary constant.

a) Let $Xu = X_3u = -\frac{1}{4}x^2u_x+uu_x+xu$. We shall solve the initial value problem

$$\begin{cases} U_t = -\frac{1}{4}x^2U_x + UU_x + xU \\ U(0,x) = -\frac{1}{4}(x+\alpha_o)^2. \end{cases}$$

The function α must satisfy the equation $\frac{d\alpha}{dt} = -\frac{1}{4}\alpha^2$ with initial condition $\alpha(0) = \alpha_o$. Therefore

$$\alpha(t) = \frac{4\alpha_o}{4+\alpha_o t} .$$

(If $\alpha_o = 0$, $\alpha(t) = 0$ for every t ; if $\alpha_o \neq 0$, $\alpha(t)$ is defined for $t \neq -4/\alpha_o$.) Therefore the solution sought is

$$U(t,x) = -\frac{1}{4}(x+\frac{4\alpha_o}{4+\alpha_o t})^2$$

b) Let $Xu = X_4u$. We solve the problem

$$\begin{cases} U_t = \frac{1}{8}x^3U_x - \frac{3}{4}x^2U - \frac{3}{2}xUU_x + U^2 \\ U(0,x) = -\frac{1}{4}(x+\alpha_o)^2. \end{cases}$$

α must satisfy $\frac{d\alpha}{dt} = -\frac{1}{8}\alpha^3$. We find that

$$U(t,x) = -\frac{1}{4}(x + \frac{2\alpha_o}{\sqrt{4+3\alpha_o^2 t}})^2.$$

For $X = X_5$ and $X = X_6$, the equation to be satisfied by α

simplifies again to $\frac{d\alpha}{dt} = -\frac{1}{32}\alpha^4$ and $\frac{d\alpha}{dt} = -\frac{1}{64}\alpha^5$, respectively.

c) Let $Xu = X_1^2 u = u_x u_{xx}$. We solve the problem

$$\begin{cases} U_t = U_x U_{xx} \\ U(0,x) = -\frac{1}{4}(x+\alpha_o)^2. \end{cases}$$

α must satisfy $\frac{d\alpha}{dt} = -\frac{1}{2}$. Therefore

$$U(t,x) = -\frac{1}{4}(x-\frac{1}{2}t+\alpha_o)^2.$$

d) Let $Xu = X_2^2 u = xu_x u_{xx} - u_x^2$. To solve this problem with the same initial condition as above, we find that α must satisfy $\frac{d\alpha}{dt} = \frac{\alpha}{2}$, whence

$$U(t,x) = -\frac{1}{4}(x+\alpha_o e^{\frac{1}{2}t})^2.$$

e) Similarly, for $Xu = X_3^2 u = \frac{1}{2}x^2 u_x u_{xx} - xu_x^2 + uu_x$, the equation for α reduces to $\frac{d\alpha}{dt} = -\frac{1}{4}\alpha^2$, and therefore the solution is the same as in a).

Example 2. If we consider the qualilinear operator $Du = u_x + u^2$ whose general solution is $\frac{1}{x-\alpha}$ (for $x \neq \alpha$), we can solve an initial-value problem with the initial condition $U(0,x) = \frac{1}{x-\alpha_o}$ for an equation such as $U_t = xU_x + U$ (since $Xu = xu_x + u$ is a symmetry of D). We recover the solution furnished by the method of characteristics ; in this case $U(t,x) = \frac{e^t}{xe^t - \alpha_o}$.

Example 3. The linear operator $Du = u_x + u$ with general solution $u(x,\alpha) = \alpha e^{-x}$ has many symmetries. We give some applications of this method to some randomly chosen problems.

a)

$$\begin{cases} U_t = U_x - U + e^x U^2 \\ U(0,x) = \alpha_o e^{-x} \end{cases}$$

We obtain the equation $\frac{d\alpha}{dt} = -2\alpha + \alpha^2$ and $\alpha = \frac{2\alpha_o}{(2-\alpha_o)e^{2t}+\alpha_o}$ for $0 < \alpha_o < 2$. The solution is

$$U(t,x) = \frac{2\alpha_o e^{-x}}{(2-\alpha_o)e^{2t}+\alpha_o} \quad .$$

b)

$$\begin{cases} U_t = e^x UU_x \\ U(0,x) = \alpha_o e^{-x} \end{cases}$$

We obtain the equation $\frac{d\alpha}{dt} = -\alpha^2$, therefore $U(t,x) = \frac{\alpha_o e^{-x}}{1+\alpha_o t}$.

c)

$$\begin{cases} U_t = 2U_x^2 + UU_x - (e^x+1)U^2 - 3e^{-x} \\ U(0,x) = \alpha_o e^{-x} \quad . \end{cases}$$

We obtain the equation $\frac{d\alpha}{dt} = -\alpha^2 - 3$, whence $\alpha = -3^{\frac{1}{2}}\tan(3^{\frac{1}{2}}t - \text{Arc}\tan(3^{-\frac{1}{2}}\alpha_o))$. The solution is

$$U(t,x) = -3^{\frac{1}{2}}\tan(3^{\frac{1}{2}}t - \text{Arc}\tan(3^{-\frac{1}{2}}\alpha_o))e^{-x}.$$

2.6. <u>Equivalence and linearization of evolution equations</u>.

We shall relate two aspects of the linearization problem, the theory of the linearization of representations of Lie algebras published in 1977 by Flato, Pinczon and Simon [11] [12] and subsequently developed with many applications (see e.g. [61] [3]), and the criterion published in the same year by Magri [43] for the equivalence of a partial differential equation of evolution

$$U_t = PU \tag{4}$$

to another equation

$$\bar{U}_t = \bar{P}\bar{U} . \tag{5}$$

This criterion is the existence of an invertible (integro-differential) operator A not depending explicitly on the variable t from the functional space of U to the functional space of $\bar{U}$ satisfying, for each solution U of (4),

$$VA(U,PU) = \bar{P}(AU). \tag{6}$$

Equivalence. We shall first study Magri's criterion in some detail. Assume first that, for each solution U of (4), (6) holds. Then A maps the solutions of (4) into solutions of (5). In fact, let U be a solution of (4) and let $\bar{U} = AU$. We obtain $\bar{U}_t = (AU)_t = VA(U,U_t) = VA(U,PU) = \bar{P}(AU) = \bar{P}\bar{U}$, proving that $\bar{U}$ is a solution of (5). If, conversely, A maps solutions of (4) into solutions of (5), then for every solution U of (4),

$$VA(U,PU) = VA(U,U_t) = (AU)_t = \bar{P}(AU)$$

which is condition (6).

We now assume that A is one-to-one on some subsets of the functional spaces and that the Fréchet derivative of A is injective. Assume first that (6) holds for each U of the form $A^{-1}\bar{U}$ where $\bar{U}$ is a solution of (5). Then A^{-1} maps the solutions of (5) into solutions of (4). In fact, assume that $\bar{U}$ is a solution of (5); set $U = A^{-1}\bar{U}$. We obtain

$$VA(U,PU) = \bar{P}(AU) = \bar{P}\bar{U} = \bar{U}_t = (AU)_t = VA(U,U_t),$$

whence $U_t = PU$ by the injectivity assumption. If, conversely, A^{-1} maps the solutions of (5) into solutions of (4), then for each solution $\bar{U}$ of (5).

$$VA(A^{-1}\bar{U},PA^{-1}\bar{U}) = VA(A^{-1}\bar{U},(A^{-1}\bar{U})_t) = \bar{U}_t = \bar{P}\bar{U}$$

which is condition (6). Therefore, under the assumption that (6) holds for each U which is a solution of (4) and for each U such that AU is a solution of (5), the equations (4) and (5) are indeed equivalent.

If, in particular, (6) holds for any function U, and if A is invertible, with injective Fréchet derivative, then A is an equivalence. In practice, P is given and one seeks A and $\bar{P}$. For polynomial operators the algorithmic methods described in 1.4 can be applied.

We note that if X and X' satisfy (1), the equation defining the symmetries of D, and if X' is independent of u, then D is an equivalence from $U_t = XU$ to $\bar{U}_t = X'\bar{U}$. We also note that if the operator A satisfies (6), the operator $B(U,\bar{U}) = \bar{U}-AU$ satisfies (3) of 1.4 with $P' = 0$ identically. Thus an equivalence is a special case of Bäcklund transformation where the equation $B(U,\bar{U}) = 0$ can be solved as $\bar{U} = AU$ and $U = A^{-1}\bar{U}$ and where $P' = 0$. (Even when B is a bidifferential operator, A or A^{-1} are in general integro-differential.)

Linearization. The main fact about equivalent equations is that all properties of one, such as the existence of symmetries, of a Hamiltonian structure, of conservation laws, and the existence of solutions to initial-value problems, are valid for the other. Since linear equations are simpler and better known than nonlinear ones, it is natural, given a nonlinear evolution equation, to look for a linear evolution equation equivalent to it, or, in effect, to linearize a given system of partial differential equations, the method proposed by Flato, et al., and also by Magri, although in apparently different terms.

Let P^1 be the linear part of the operator P, i.e., the coefficient of λ in the expansion of $P(\lambda U)$ in powers of λ. Up to a linear transformation, to linearize $U_t = PU$ is to solve the equation for the operator A from the functional space $\mathcal{U}$ of U to itself,

$$VA \circ P = P^1 \circ A. \tag{7}$$

Given a nonlinear representation [11] ρ of a Lie algebra $\mathcal{g}$ in the Lie algebra of nonlinear operators on a topological space $\mathcal{U}$, to linearize the representation is, by definition, to find an operator on $\mathcal{U}$ which linearizes all the operators P in $\rho(\mathcal{g})$.

In [11] and in the subsequent papers, the condition for the formal linearization of a representation is expressed in terms of star-products of formal power series as $A{*}P = P^1 \circ A$ for each $P \in \rho(\mathcal{g})$. It is a simple observation that for an integro-differential operator A expanded in formal power series of multilinear operators on $\mathcal{U}$, $A{*}P = VA \circ P$. Thus the formal linearization problem for a one-dimensional Lie algebra, in the sense of [11], coincides with the search for an equivalent linear equation in the sense of [43]. Of course, it is not only the formal linearization problem which is of interest, and much work has been devoted to proving that the formal power series first discovered by solving the equation $A{*}P = P^1 \circ A$ converges. (See, e.g., [62][61].)

Example 1. We return to Burgers's equation and we show, following Magri [43], that the operator A_o defined by $A_o U = e^{\frac{1}{2}\int U}$, where $\int U = \int_{-\infty}^{x} U(t,\xi)d\xi$ and whose inverse is $B_o\bar{U} = A_o^{-1}\bar{U} = 2\frac{\bar{U}_x}{\bar{U}}$, the Hopf-Cole transformation, is an equivalence from Burgers's equation, $U_t = UU_x + U_{xx}$, into the heat equation, $\bar{U}_t = \bar{U}_{xx}$, and therefore linearizes Burgers's equation.

We set $PU = UU_x + U_{xx}$ and $P^1\bar{U} = \bar{U}_{xx} = \bar{P}\bar{U}$. We calculate

$$VB_o(\bar{U},\bar{P}\bar{U}) = \frac{2}{\bar{U}}\,\bar{U}_{xxx} - \frac{2\bar{U}_x}{\bar{U}^2}\,\bar{U}_{xx} \text{ and}$$

$$P(B_o\bar{U}) = \frac{4\bar{U}_x}{\bar{U}}\left(\frac{\bar{U}_{xx}}{\bar{U}} - \frac{\bar{U}_x^2}{\bar{U}^2}\right) + 2\left(\frac{\bar{U}_{xxx}}{\bar{U}} - \frac{3\bar{U}_x\bar{U}_{xx}}{\bar{U}^2} + \frac{2\bar{U}_x^3}{\bar{U}^3}\right)$$

$$= \frac{2}{\bar{U}}\,\bar{U}_{xxx} - \frac{2\bar{U}_x}{\bar{U}^2}\,\bar{U}_{xx}\,.$$

We have thus proved that $VB_o(\bar{U},\bar{P}\bar{U}) = P(B_o\bar{U})$ for any function $\bar{U}$ which does not vanish, thereby proving that (6) holds for any inverse A_o of B_o identically in U. Thus A_o linearizes Burgers's equation into the heat equation.

Example 2. In [62] Taflin proved, among other interesting results, that Burgers's equation is linearizable into the heat equation by another integral operator defined by $AU = Ue^{\frac{1}{2}\int U}$. To verify this fact by the method of Fréchet derivatives (Magri's method), it is enough to compute VA(U,PU) and $\bar{P}(AU)$ and to note that both are equal to $(U_{xx} + \frac{1}{4}U^3 + \frac{3}{2}UU_x)e^{\frac{1}{2}\int U}$ for any function U such that the integral $\int U$ exists. The operator A has an inverse, $A^{-1}\bar{U} = \dfrac{\bar{U}}{1 + \frac{1}{2}\int\bar{U}}$, which is defined on the space of rapidly decreasing functions with $1 + \frac{1}{2}\int\bar{U} > 0$.

Application to the determination of symmetries.

For a linear operator $\bar{X}$, the operator $\bar{X}$ itself or any linear symmetry of it constitues a recursion operator, and when $\bar{U}_t = \bar{X}\bar{U}$ possesses a symmetry, it possesses an infinite sequence of generalized symmetries obtained by successive applications of the recursion operator. If a nonlinear operator can be linearized into a linear operator $\bar{X}$ having a sequence $\bar{X}_n$ of commuting generalized symmetries, X has a sequence of commuting generalized symmetries X_n obtained by $X_n U = VA^{-1}(U,\bar{X}_n(AU))$. We return to the example of Burgers's equation. Since the heat equation into which it can be linearized by either of the two operators A_o or A of examples 1 and 2 possesses the sequence of commuting generalized symmetries $\bar{X}_n U = \frac{\partial^n U}{\partial x^n}$, a corresponding sequence of commuting symmetries X_n of Burgers's equation can be derived by means of the linearization operators A_o or A (The results of these computations are also to be found in [44] (6.2.) and in [62] (9) respectively.) We obtain a convenient way of computing a sequence of commuting generalized symmetries of Burgers's equation which turn out to be the generalized symmetries produced by Olver's recursion operator [49].

$$\bar{X}_n(AU) = \frac{\partial^n}{\partial x^n}\left(Ue^{\frac{1}{2}\int U}\right) = S_n(U)\, e^{\frac{1}{2}\int U}$$

with $S_o(U) = U$, $S_1(U) = U_x + \frac{1}{2}U^2$, $S_2(U) = U_{xx} + \frac{3}{2}UU_x + \frac{1}{4}U^3, \ldots$, and

$$S_n(U) = (S_{n-1}(U))_x + \frac{1}{2}US_{n-1}(U).$$

Also $VA^{-1}(\bar{U},\bar{V}) = \dfrac{1}{\left(1+\frac{1}{2}\int\bar{U}\right)^2}\left(\bar{V} + \frac{1}{2}\bar{V}\int\bar{U} - \frac{1}{2}\bar{U}\int\bar{V}\right)$, therefore

$$VA^{-1}(AU,\bar{V}) = e^{-\int U}\left(\bar{V}e^{\frac{1}{2}\int U} - \frac{1}{2}Ue^{\frac{1}{2}\int U}\int\bar{V}\right) = e^{-\frac{1}{2}\int U}\left(\bar{V} - \frac{1}{2}U\int\bar{V}\right).$$

Hence $X_n U = VA^{-1}(AU, \bar{X}_n(AU)) = e^{-\frac{1}{2}\int U}(\bar{X}_n(AU) - \frac{1}{2}U\int \bar{X}_n(AU))$

$$= e^{-\frac{1}{2}\int U}(\bar{X}_n(AU) - \frac{1}{2}U\bar{X}_{n-1}(AU))$$

$$= S_n(U) - \frac{1}{2}US_{n-1}(U)$$

$$= (S_{n-1}(U))_x \ .$$

We have thus obtained the following formula

$$X_n U = (S_{n-1}(U))_x = \frac{\partial}{\partial x}(e^{-\frac{1}{2}\int U} \frac{\partial^{n-1}}{\partial x^{n-1}}(Ue^{\frac{1}{2}\int U})).$$

The calculation of the first terms in the sequence yields
$X_1 U = U_x$, $X_2 U = U_{xx} + UU_x$, $X_3 U = U_{xxx} + \frac{3}{2}UU_{xx} + \frac{3}{2}U_x^2 + \frac{3}{4}U^2 U_x$,

which are indeed the symmetries $B^{(1)}$, $B^{(2)}$, $B^{(3)}$,... found in [49]. If we carry out similar computations using the linearization operator $A_o U = e^{\frac{1}{2}\int U}$ of example 1 we obtain the same symmetries X_n, up to a multiplication factor of 2.

Note that the linearization of Burgers's equation by means of the Hopf-Cole transformation is also derived in [25], as a corollary of the linearization of all the equations $u_t = u_{xx} + \alpha(u)u_x^2$.

Conclusion.

The study of the symmetries of systems of partial differential equations is an active field with important consequences for theoretical physics. The problems that remain to be solved are numerous in particular the explicit determination of the Lie algebra of symmetries for examples other than the familiar [67] Burgers and Korteweg-de Vries equations, general theorems on the structure of the Lie algebra of systems of partial differential equations -

results hage been obtained for classical symmetries by Ovsiannikov [54] and by Olver [50] for linear equations -, the comparison of the Lie algebras of equivalent systems of equations, the clarification of the relationships among several properties shared by some equations, such as the existence of an infinity of commuting symmetries, the existence of two Hamiltonian structures, the solvability by the spectral transform (inverse scattering), the existence of auto-Bäcklund transformations, and linearizability. Different approaches are possible (see e.g. [45] [57] [15]); many problems remain open.

References

[1] Anderson, R.L. and Ibragimov, N.H., Lie-Bäcklund Transformations in Applications, SIAM Studies in Applied Mathematics 1, 1979.

[2] Anderson, R.L., Kumei, S. and Wulfman, C.E., Generalization of the concept of invariance of differential equations, Phys. Rev. Lett. 28 (1972) pp. 988-991.

[3] Anderson, R.L. and Taflin, E., Explicit non-soliton solutions of the Benjamin - Ono equations, Lett. Math. Phys. 7 (1983) pp. 243-248.

[4] Benjamin, T.B. and Olver, P.J., Hamiltonian structure, symmetries and conservation laws for water waves, J. Fluid Mech. 125 (1982) pp. 137-185.

[5] Bessel-Hagen, E., Uber die Erhaltungssätze der Elektrodynamik, Math. Ann. 84 (1921) pp. 258-276.

[6] Bluman, G.W. and Cole, J.D., Similarity Methods for Differential Equations, Springer-Verlag, 1974.

[7] Branson, T.P., Conformally invariant equations on differential forms, Comm. Partial. Diff. Eq. 7 (1982) pp. 393-431.

[8] Branson, T.P., Intertwining differential operators for spinor-form representations of conformal groups, preprint, 1982.

[9] Branson, T.P. and Kosmann-Schwarzbach, Y., Conformally covariant nonlinear equations on tensor-spinors, Lett. Math. Phys. 7 (1983) pp. 63-73.

[10] Edelen, D.G.B., Isovector Methods for Equations of Balance, Sijthoff and Noordhoff, 1980.

[11] Flato, M., Pinczon, G., and Simon, J., Nonlinear representations of Lie groups, Ann. Sci. Ec. Norm. Sup. Paris, Sér. IV, 10 (1977) pp. 405-418.

[12] Flato, M., and Simon, J., Nonlinear equations and covariance, Lett. Math. Phys. 2 (1977) pp. 155-160.

[13] Fokas, A.S., Invariants, Lie-Bäcklund operators and Bäcklund transformations, Ph. D. thesis, California Institute of Technology, 1979.

[14] Fokas, A.S., A symmetry approach to exactly solvable evolution equations, J. Math. Phys. 21 (6) (1980), pp. 1318-1325.

[15] Fokas, A.S. and Anderson, R.L., On the use of isospectral eigenvalue problems for obtaining hereditary symmetries for Hamiltonian systems, J. Math. Phys. 23 (6) (1982) pp. 1066-1073.

[16] Fokas, A.S. and Fuchssteiner , B., Bäcklund transformations for hereditary symmetries, Nonlinear Anal., Theory, Methods and Appl. 5 (4) (1981) pp. 423-432.

[17] Fokas, A.S. and Fuchssteiner B., The hierarchy of the Benjamin-Ono equation, Phys. Lett. 86A (6,7) (1981) pp. 341-345.

[18] Fuchssteiner, B., Application of hereditary symmetries to nonlinear evolution equations, Nonlinear Anal., Theory, Methods and Appl. 3 (1979) pp. 849-862.

[19] Fushchich, V.I. and Nikitin, A.G., On the new invariance algebras of relativistic equations for massless particles, J. Phys. A 12 (6) (1979) pp. 747-757.

[20] Gelfand, I.M. and Dikii, L.A., Asymptotic behaviour of the resolvent of Sturm-Liouville equations and the algebra of the Korteweg-de Vries equations, Usp. Mat. Nauk 30 (5) (1975) pp. 67-100 ; Engl. transl., Russian Math. Surveys 30 (5) (1975) pp. 77-113.

[21] Gragert, P.K.H., Kersten, P.H.M. and Martini, R., Symbolic computations in applied differential geometry, Acta Appl. Math. 1 (1) (1983) pp. 43-77.

[22] Gürsey, F., On a conform-invariant spinor wave equation, Nuovo Cimento 3 (1956) pp. 988-1006.

[23] Hermann, R., E. Cartan's geometric theory of partial differential equations, Advances in Math. 1 (1965) pp. 265-317.

[24] Hermann, R., Geometry, Physics and Systems, Mareel Dekker 1973.

[25] Ibragimov, N. Kh. and Shabat, A.B., Evolutionary equations with nontrivial Lie-Bäcklund group, Funk. Anal. Pril. 14 (1) (1980) pp. 25-36 ; Engl. transl. Funct. Anal. Appl. 14 (1980) pp. 19-28.

[26] Johnson, H.H., Bracket and exponential for a new type of vector field, Proc. Amer. Math. Soc. 15 (1964) pp. 432-437.

[27] Johnson, H.H., A new type of vector field and invariant differential systems, Proc. Amer. Math. Soc. 15 (1964) pp. 675-678.

[28] Kolar, I., On the second tangent bundle and generalized Lie derivatives, Tensor N.S. 38 (1982) pp. 98-102.

[29] Konopelchenko, B.G. and Mokhnachev, V.G., On the group theoretical analysis of differential equations, Yad. Fiz. 30 (1979) pp. 559-567 (Engl. transl., Sov. J. Nucl. Phys. 30 (2) 1979) pp. 288-292) and J. Phys. A13 (1980) pp. 3113-3124.

[30] Kosmann-Schwarzbach, Y., Sur les transformations de similitude des équations aux dérivées partielles, C.R. Acad. Sc. Paris 287 A (1978) pp. 953-956.

[31] Kosmann-Schwarzbach, Y., Generalized symmetries of nonlinear partial differential equations, Lett. Math. Phys. 3 (1979) pp. 395-404.

[32] Kosmann-Schwarzbach, Y., Vector fields and generalized vector fields on fibered manifolds, Lect. Notes Math. 792, Springer-Verlag, 1980, pp. 307-355.

[33] Kosmann-Schwarzbach, Y., Hamiltonian systems on fibered manifolds, Lett. Math. Phys. 5 (1981) pp. 229-237.

[34] Kruskal, M., Miura, R., Gardner, C. and Zabusky, N., Korteweg-de Vries equation and generalizations V, J. Math. Phys. 11 (1970) pp. 952-960.

[35] Kumei, S., Invariance transformations, invariance group transformations and invariance groups of the sine-Gordon equations, J. Math. Phys. 16 (12) (1975) pp. 2461-2468.

[36] Kumei, S., On the relationship between conservation laws and invariance groups of nonlinear field equations in Hamilton's canonical form, J. Math. Phys. 19 (1) (1978) pp. 195-199.

[37] Kupershmidt, B.A., Geometry of jet bundles and the structure of Lagrangian and Hamiltonian formalisms, Lecture Notes Math. 775, Springer-Verlag, 1980, pp. 162-218.

[38] Lax, P.D., A Hamiltonian approach to the KdV and other equations, in Nonlinear Evolution Equations, Academic Press, 1978, pp. 207-224.

[39] Lefebvre, J.Y. and Metzger, P., Quelques exemples de groupes d'invariance d'équations aux dérivées partielles, C.R. Acad. Sc. Paris, 279A (1974) pp. 165-168.

[40] Lie, S., Gesammelte Abhandlungen, Vol. 3, Teubner, Leipzig, 1922.

[41] Magri, F., An operator approach to Poisson brackets, Ann. Phys. (N.Y.) 99 (1976), pp. 196-228.

[42] Magri, F., An operator approach to symmetries, Nuovo Cimento 34B (2) (1976), pp. 334-344.

[43] Magri, F., Equivalence transformations for nonlinear evolution equations, J. Math. Phys. 18 (7) (1977), pp. 1405-1411.

[44] Magri, F., Properties of the Hamiltonian equations reducible to linear equations, Proc. Joint IUTAM/IMU Symposium, Novosibirsk, 1978.

[45] Magri, F., A geometrical approach to the nonlinear solvable equations, Lecture Notes Phys. 120, Springer-Verlag, 1980, pp. 233-263.

[46] Manin, Yu.I., Algebraic aspects of nonlinear differential equations, Itogi Nauki i Tekhniki 11 (1978) pp. 5-152 ; Engl. transl., J. Soviet Math. 11 (1979) pp. 1-122.

[47] Metzger, P., Quelques autres exemples de groupes d'invariance d'équations aux dérivées partielles, C.R. Acad. Sc. Paris 279A (1974) pp. 193-196.

[48] Oevel, W. and Fokas, A.S., Infinitely many commuting symmetries and constants of motion in involution for explicitely time-dependent evolution equations, preprint, 1981.

[49] Olver, P.J., Evolution equations possessing infinitely many symmetries, J. Math. Phys. 18 (6) (1977) pp. 1212-1215.

[50] Olver, P.J., On the symmetry group of a linear partial differential equation, preprint, 1978.

[51] Olver, P.J., How to find the symmetry group of a differential equation, Appendix in Sattinger, D.H., Group Theoretic Methods in Bifurcation Theory, Lecture Notes Math. 762, Springer-Verlag, 1979.

[52] Olver, P.J., Applications of Lie Groups to Differential Equations, Lecture Notes, Mathematical Institute, Oxford 1980.

[53] Olver, P.J., On the Hamiltonian structure of evolution equations, Math. Proc. Camb. Phil. Soc. 88 (1980) pp. 71-88.

[54] Ovsiannikov, L.V., Group Analysis of Differential Equations, Academic Press, 1982. (Russian edition, Nauka, Moscow, 1978.)

[55] Paneitz, S.M. and Segal, I.E., Analysis in space-time bundles, I. General considerations and the scalar bundle, J. Funct. Anal. 47 (1982) pp. 78-142, and II. The spinor and form bundles, ibid. 49 (1982) pp. 335-414.

[56] Rosenau, P. and Schwarzmeier, J.L., Similarity solutions of systems of partial differential equations using MACSYMA, Courant Inst., Magneto-fluid dynamics division, 1979.

[57] Samohin, A.V., Symmetries of Sturm-Liouville equations and the Korteweg-de Vries equation, Dokl. Akad. Nauk S.S.S.R. 251 (3) (1980) pp. 557-561 ; Engl. transl., Soviet Math. Dokl. 21 (2) (1980) pp. 488-492.

[58] Samohin, A.V., On symmetries of linearizable evolution systems of equations, Dokl. Akad. Nauk. S.S.S.R. 262 (2) (1982), pp. 274-279 ; Engl. transl., Soviet Math. Dokl. 25 (1) (1982) pp. 56-61.

[59] Schwarz, F., Symmetries of SU (2) invariant Yang-Mills theories, Lett. Math. Phys. 6 (1982) pp. 355-359.

[60] Segal, I.E., La variété des solutions d'une équation hyperbolique, non linéaire d'ordre 2, Collège de France, Paris 1964-1965.

[61] Simon, J.C.H., Global solutions for the coupled Maxwell-Dirac equations, Lett. Math. Mhys. 6 (1982) pp. 487-489.

[62] Taflin E., Analytic linearization, Hamiltonian formalism and infinite sequences of constants of motion for the Burgers equation, Phys. Rev. Lett. 47 (20) (1981) pp. 1425-1428.

[63] Tu, G.Z., On the similarity solution of evolution equation $u_t = H(x,t,u,u_x,u_{xx},...)$, Lett. Math. Phys. 4 (1980) pp. 347-355.

[64] Vinogradov, A.M., The theory of higher infinitesimal symmetries of nonlinear partial differential equations, Dokl. Akad.Nauk SSSR 248 (2) (1979) pp. 274-278 ; Engl. transl. Soviet Math. Dokl. 20 (5) (1979) pp. 985-990.

[65] Vinogradov, A.M., Geometry of nonlinear differential equations, Itogi Nauki i Tekhniki, Seriya Problemy Geometrii, 11 (1980) pp. 89-134 ; Engl. transl., J. Soviet Math. 11 (1980) pp. 1624-1649.

[66] Vinogradov, A.M., Equations on Manifolds, New Developments in Global Analysis, Voronezh, 1981.

[67] Vinogradov, A.M. and Krasil'ščik, I.S., A method of computing higher symmetries of nonlinear evolution equations, and non-local symmetries, Dokl. Akad. Nauk SSSR 253 (6) (1980) pp. 1289-1292 : Engl. transl., Soviet Math. Dokl. 22 (1) (1980) pp. 235-239.

A LIE ALGEBRAIC APPROACH TO ORDER PARAMETERS

Allan I. Solomon

Faculty of Mathematics
The Open University
Milton Keynes, U.K.

Abstract

We consider physical systems which exhibit phase transitions. In many exactly-solvable cases, the condensed phase is described by a Hamiltonian which is a representative of an element of a Lie algebra. When the phase transition breaks an abelian symmetry, it is shown that order parameters are given by the generators of the one-dimensional root spaces of a Cartan basis.

1. LIE ALGEBRAS AND PHASE TRANSITIONS

In statistical physics, the thermodynamic properties of a system described by a hamiltonian H - a self-adjoint operator in Hilbert space - are derived from the partition function

$$Z(\beta) = \text{trace}\ (\exp(-\beta H)) \tag{1}$$

The temperature effects are supposedly contained in the single, real positive parameter $\beta = 1/kT$, where k is Boltzmann's constant and T is the absolute temperature. In many exactly-solvable models, the hamiltonian H is an element of (a representation) of a semi-simple Lie algebra g, the exponent a member of the corresponding Lie group representation, and so it comes as no surprise to find that elementary Lie groups analysis is of value in the elucidation of the thermodynamic properties of such systems. The thermodynamic average of an observable ϕ is given by

$$\langle\phi\rangle_{H,\beta} = \text{trace}\ (\exp(-\beta H)\phi)/Z(\beta) \tag{2}$$

S. Sternberg (ed.), Differential Geometric Methods in Mathematical Physics, 279–285.

Again, in examples of physical interest, ϕ is an element of (the representation of) g. Although such calculations are manifestly representation-dependent, Lie algebraic criteria can be given to ensure, for example, the vanishing of $\langle\phi\rangle_{H,\beta}$ in any representation. In particular, we shall be interested in systems exhibiting phase transitions. These may be described in terms of order parameters, operators ϕ whose vanishing (resp. non-vanishing) indicates the absence (resp. presence) of a particular phase. Examples of order parameters are the magnetization of a ferromagent, or the electron-pair operator in a superconductor.

Phase transitions usually manifest themselves by the spontaneous breaking of a symmetry group[1]. An approach to this phenomenon is to postulate that the system above the transition temperature T_C is described by a hamiltonian H_I which is invariant under the symmetry group: in general, such a hamiltonian is not an element of the Lie algebra g, but commutes with a Hilbert space representation of the Lie algebra g_S of the symmetry group. Order parameters for the phase must vanish in eigenstates of H_I. Below the transition temperature T_C, we describe the system by means of a hamiltonian H, as in the opening remarks, which <u>is</u> an element of g, but which does not commute with g_S, and for which the order parameters do not vanish. The algebra g is not a symmetry algebra of H, but plays a role akin to that of the spectrum-generating algebras of particle physics; it is generated by the operators of interest in the model, including the hamiltonian H, the order parameters, and the relevant conserved quantities (elements of g_S), and its finite-dimensionality leads to the exact solvability of the system. We shall only consider the case where the algebra g_S is abelian; and we take g_S to be contained in the Cartan subalgebra h of g:

$$g_S \subset h \subset g.$$

In these circumstances it is straightforward to characterize a set of operators which play the role of order parameters; they are the elements of the root spaces of g. Recall the definition of a root of (g,h); given $\lambda \in h^*$, the dual space of h, define the set

$$g_\lambda = \{y \in g: [x,y] = \lambda(x)y \quad \text{for all } x \in h\}.$$

Then λ is called a root if $\lambda \neq 0$ and $g_\lambda \neq \{0\}$. The Hilbert space representation $\hat{y}$ of the abstract Lie algebra element y vanishes in eigenstates of H_I. For, since H_I commutes with $\hat{h}_i$ by hypothesis, we may label the eigenstates of H_I by labels $n_1, n_2, \ldots, n_\ell$ (corresponding to eigenvalues of the $\hat{h}_i$) and other labels ξ; the equation in conventional notation for eigenvalues $E(n_i;\xi)$ being

$$H_I|n_i;\xi> = E(n_i;\xi)|n_i;\xi>.$$

For any $y \in g_\lambda, h_k \in h$

$$[h_k, y] = \lambda(h_k)y;$$

In Hilbert space

$$[\hat{h}_k, \hat{y}] = \lambda(h_k)\hat{y},$$

and, since

$$<n_i;\xi|[\hat{h}_k, \hat{y}]|n_i;\xi> = 0,$$

we have

$$<n_i;\xi|\hat{y}|n_i;\xi> = 0$$

because $\lambda(h_k)$ is a non-zero number for at least one k. Specifically, $\hat{y}$ acts as an order parameter for the phase corresponding to the breaking of the symmetry generated by $\hat{h}_k$ (non-conservation of $\hat{h}_k$). The root spaces are one-dimensional and a standard Cartan basis for the n-dimensional rank-ℓ Lie algebra g may be chosen as

$$\{h_1, h_2, \ldots, h_\ell;\ e_1, e_2, \ldots, e_{n-\ell}\}$$

where the commuting h_i form a basis for the ℓ-dimensional Cartan subalgebra h, and the e_i are the basis vectors for the root spaces. We have n-ℓ independent order parameters $\hat{e}_i$. Since H is a linear combination of $\hat{h}_i$ and $\hat{e}_j$, in general $\hat{e}_j$ will not vanish in eigenstates of H.

The algebraic solution to the physical system described by a diagonalizable hamiltonian H proceeds as follows: H is a representative of a semi-simple element v of g (since it is diagonalizable) and so there exists an (inner) automorphism $\phi: g \to g$ such that

$$v \to \phi(v) = \alpha_1 h_1 + \ldots + \alpha_\ell h_\ell \in h \qquad \alpha_i \in \mathbb{C}$$

This ϕ may be implemented in any faithful representation of g, in which v is represented by some matrix M. The α_i are then determined as functions of ℓ invariants of the form tr M^n. The Hilbert space implementation

$$H \to UHU^{-1} = \alpha_1\hat{h}_1 + \ldots + \alpha_\ell\hat{h}_\ell$$

is known as the Bogoliubov transformation[2] in the literature.

Given the spectra of the $\hat{h}_i$, the spectrum of H follows immediately; this is the spectrum of the physical system.

The thermodynamic behaviour follows similarly: from (1)

$$Z(\beta) = \text{trace}(U\exp(-\beta H)U^{-1})$$

$$= \text{trace}\exp(-\beta\sum\alpha_i\hat{h}_i)$$

$$= \sum_{\underline{n}}\exp(-\beta\sum\alpha_i n_i)$$

where $\underline{n} = (n_1, n_2, \ldots, n_\ell)$ is a simultaneous eigenvalue of the $\hat{h}_i$. The thermodynamic expectation of an order parameter $\hat{e}_j$

$$\langle\hat{e}_j\rangle_{H,\beta} = \text{trace}\{(\exp-\beta H)\hat{e}_j\}/Z(\beta)$$

$$= \text{trace}\{\exp(-\beta\sum^{\ell}\alpha_i\hat{h}_i)U\hat{e}_jU^{-1}\}/Z(\beta)$$

$$= \text{trace}\exp(-\beta\sum_1^{\ell}\alpha_i\hat{h}_i)\sum_1^{\ell}\alpha_{jk}\hat{h}_k\}/Z(\beta)$$

The last line follows because the terms in $U\hat{e}_jU^{-1}$ involving $\hat{e}_i$'s vanish in eigenstates of the $\hat{h}_i$. Thus

$$\langle\hat{e}_j\rangle_{H,\beta} = \sum_{\underline{n}}\{\exp(-\beta\sum_1^{\ell}\alpha_i n_i)\sum_1^{\ell}\alpha_{jk}n_k\}/Z(\beta)$$

We next illustrate these ideas in a model which is mathematically trivial, but physically rather important.

2. SL(2,C) Model - Superfluidity and Superconductivity

The simplest illustration of the discussion of the previous section is the Lie algebra $s\ell(2,C)$ - two real forms of which correspond to the two physical examples of a superconducting (fermion) system and a superfluid (boson) system.

We choose the following (Cartan) basis:

$$x = \begin{bmatrix} 0 & 1 \\ 0 & 0 \end{bmatrix} \quad y = \begin{bmatrix} 0 & 0 \\ 1 & 0 \end{bmatrix} \quad h = \begin{bmatrix} 1 & 0 \\ 0 & -1 \end{bmatrix} \qquad (3)$$

and the commutation relations are:

$$[x,y] = h \qquad [h,x] = 2x \qquad [h,y] = -2y$$

In this model there is only one conserved quantity, corresponding to h in this representation. The symmetry algebra of the hamiltonian H_T above the transition temperature T_C is that generated by the element $\hat{h}$, the Hilbert space representative of h. In the two physical systems described by SL(2,C), $\hat{h}$ is essentially the number operator N, and the phase transition corresponds to the spontaneous breaking of N-conservation, or, equivalently, gauge invariance. The operators $\hat{x}$ and $\hat{y}$ vanish in eigenstates of $\hat{h}$, and are thus the order parameters in this model. In the usual physics treatments, the hamiltonian representing the condensed phase (mean-field hamiltonian) is a Hilbert space representative of a general element v of $s\ell(2,C)$:

$$v = ax + by + ch = \begin{bmatrix} c & a \\ b & -c \end{bmatrix} \tag{4}$$

By hypothesis the hamiltonian (4) does not conserve $\hat{h}$,

$$[h,v] = 2ax - 2by \neq 0$$

and so one at least of a and b is non-zero. It is of interest to evaluate the thermodynamic expectation (2) of an element $w = a'x + b'y + c'h$ in this representation:

$$\langle w\rangle_{v,\beta} = \frac{-k(v,w)}{k(v,v)^{\frac{1}{2}}} \tanh \beta k(v,v)^{\frac{1}{2}} \tag{5}$$

Here k(v,w) is the invariant bilinear form associated with the representation (3)

$$k(v,w) = \frac{1}{2}\, \mathrm{tr} vw$$

which corresponds to the Killing form for the adjoint representation. The result (5), although representation-dependent, exhibits the characteristic tanh behaviour of mean-field theories, and is, significantly, a function only of algebraic invariants. Note that k(v,w) is non-zero for at least one of the order parameters w = x or w = y.

The excitation spectrum for such a system is also of physical interest, and even more immediate, being simply $k(v,v)^{\frac{1}{2}} = (\mathrm{tr} v^2)^{\frac{1}{2}}$, times the eigenvalues of h.

As remarked, the physical systems of superconductivity and superfluidity correspond to real forms of SL(2,C). For superconductivity, the hamiltonian for a paired fermion system in the mean-field approximation is represented by

$$v = \varepsilon h + \Delta x + \Delta^* y$$

(actually, a direct sum of such terms indexed by momentum - but this doesn't change the general discussion) where ε is the energy (real), and Δ is a complex number. The elements $\{\frac{1}{2}(x + y), -\frac{i}{2}(x - y), \frac{1}{2}h\}$ generate so(3) - which is the Lie algebra associated with this system. The excitation spectrum is determined by

$$k(v,v)^{\frac{1}{2}} = (\varepsilon^2 + |\Delta|^2)^{\frac{1}{2}}$$

and equation (5) for the order parameter y, say, is

$$\langle y \rangle_{v,\rho} = \tfrac{1}{2}\Delta \frac{\tanh \beta(\varepsilon^2 + |\Delta|^2)^{\frac{1}{2}}}{(\varepsilon^2 + \Delta^2)^{\frac{1}{2}}}$$

In the usual treatment[3] $\langle y \rangle_{v,\beta}$ is taken equal to Δ, which leads to a self-consistent determination of Δ.

For a superfluid boson system[4] interacting with real potential V, the mean field hamiltonian is represented by

$$v = (\varepsilon + N_0 V)h + N_0 V(x - y)$$

where N_0 is the density of superfluid condensate. The elements $\{\frac{1}{2}(x - y), -\frac{i}{2}(x + y), \frac{1}{2}h\}$ generate so(2,1) - which is the spectrum-generating algebra associated with this system. The excitation spectrum is determined by

$$k(v,v)^{\frac{1}{2}} = (\varepsilon^2 + 2N_0 V\varepsilon)^{\frac{1}{2}}$$

and the self-consistent order parameter equation is

$$\langle x \rangle_{v,\beta} = \tfrac{1}{2}N_0 V \frac{\tanh \beta(\varepsilon^2 + 2N_0 V\varepsilon)^{\frac{1}{2}}}{(\varepsilon^2 + 2N_0 V\varepsilon)^{\frac{1}{2}}}$$

where $\langle x \rangle_{v,\beta}$ is taken to be the superfluid condensate.

The above calculations were performed in the lowest 2×2 representation of SL(2,C) for illustrative purposes. Evaluation of the physical order parameters should be carried out in Hilbert space - as at the end of the first section. But the <u>form</u> of the spectrum and the order parameter equations is maintained in every representation.

Although the Lie algebraic approach to statistical mechanical problems enables elegant solutions to be obtained which would nonetheless be readily derivable by other methods, the value of the method becomes obvious in more complex systems involving several simultaneous phases[5].

References

1. Michel, L. 1980. Rev. Mod. Phys. 52, pp.617-651
2. Bogoliubov, N.N. 1947. J. Phys. (USSR) 11, p.23
3. Bardeen, J., Cooper, L.N. and Schrieffer, J.R. 1957 Phys. Rev. 108, p.1125
4. Solomon, A.I. 1971. J.Math.Phys.12,pp.390-394
5. Birman, J.L. and Solomon, A.I. 1982. Phys. Rev. Lett. 49,3, p.230

INDEX